Deepen Your Mind

Deepen Your Mind

前言

2018 年，我有很長一段時間在中國和美國兩地跑，同時在中國大陸工作和生活了比較長的一段時間，這是我近二十年來第一次和中國大陸的開發者一起長時間工作。在享受各種美食之外，對中國大陸的開發、產品和管理有了全新的了解和認識。

説起寫書的緣由，我本來的想法只是寫一點可以作為國內工程師教育訓練教材的東西。2018 年初，TensorFlow 作為一個技術熱點，逐漸普及到機器學習應用程式開發的各方面，但是對於 TensorFlow 在行動端的開發和應用還處於初始階段。我當時也剛剛結束一個 TensorFlow 專案，想把這些經驗和想法沉澱一下。於是我就把以前寫的筆記和記錄檔重新整理，增加一些內容並修改了文字，基本形成了一個原始版本。

後來，遇到博文視點的南海寶編輯，透過商談，出版社欣然同意把這些資料整理出書。我的筆記和記錄檔的內容很多和程式緊密相關，其中很多內容後來演變成了文件，我覺得這對初學者和有經驗的開發者都是一個很好的參考，至少可以提供另外一個角度，讓開發者多方面了解 TensorFlow。所以，我就開始寫作，前後花費了近兩年的時間。

我是一邊寫作一邊工作的，在這過程中很快就遇到了兩個很大的挑戰。

第一是文字。我的筆記都是英文的，要把這些轉換成中文，我借助了 Google 翻譯，雖然翻譯後的文字有很多需要修改，但至少省下了不少打字的時間。另外，就是專有術語的翻譯，由於我對中文的專業術語不熟悉，所以即使簡單的術語也要斟酌確定，這也花費了一些時間。如果讀者在文字中發現一些奇怪的說法，還請見諒，我和編輯雖然盡了最大的努力，可能還是會有很多遺漏。

第二是重新認識和了解了中國大陸開發的各方面。我在美國和中國大陸的開發者也有不少接觸，我想在兩邊工作應該不會有什麼差別，可實際工作起來還是有很多不同和挑戰，感觸頗深。首先是技術層面。開放原始碼的理念和軟體在中國大陸滲透到各方面，幾乎所有網際網路公司都是從使用開放原始碼軟體開始架設自己的產品。由於 Google 在開放原始碼社群的貢獻和影響力，中國大陸普遍對 Google 的好感度很高，我也同享了這個榮耀。而且，很多公司和開發者也把對開放原始碼社群做出貢獻看作責任和榮耀，這是一個很好的趨勢，中國很快會發展出自己的開放原始碼生態和社群。

關於開發環境和工程師文化，我想提一下兩邊對新員工教育訓練的區別。在中國大陸對新員工的教育訓練中，職業道德教育訓練和公司文化的教育訓練佔了很大一部分。而在矽谷，至少像 Google、臉書這些公司，教育訓練中技術教育訓練佔了很大一部分，基本是一周的教育訓練後，員工就要進行實際的工作，而中國大陸很多公司的新員工第二周才開始技術工作。這裡我能充分感受到中美公司之間的差別。

另外是開發管理方法，由於管理方法的不同，實際的工作中要做對應的改變。例如中國大陸對開發和產品的進度的管理是非常嚴格的。但是，這種嚴格大都表現在層級的匯報關係上，而非對技術細節的掌控和指導上。Google 的工程師會經常以程式的提交作為一個專案開始和結束的標示，這在中國大陸公司很少見到。

我希望把這些經驗、想法和體會能或多或少表現在這本書裡。舉例來說，使用 Markdown 寫文件，能使寫文件變成一件不是很煩瑣的事，可以讓作者更專注於內容的寫作，而非花費太多時間在操作編輯器上。本書就是全部用 Markdown 寫作完成，再轉換成 Word 文件的。舉例來說，使用 Bazel 編譯，需要對程式的依賴有清晰的定義。可能很多工程師不會特別在意這點，但是透過它，工程師可以非常清楚地了解程式重用和參考的狀況，避免隨意的程式重用，並加強程式的品質。我希望透過這些在書中傳達給讀者一些不同的開發經驗。

總之，我會把本書作為這一年工作和生活的紀念。看到書中的各個章節，我就可以聯想起寫書時發生的許多事。但是，真的由於時間和我自己的能力非常有限，書中一定會有很多錯誤和瑕疵，還望讀者能寬容和諒解。

最後，要感謝我的家人能支持和陪伴我度過 2018 年，我和我的母親一起度過了 2018 年春節，是近 20 年來在中國大陸度過的第一個春節。還要感謝我的妻子，她非常支持我，並幫助我寫完這本書。還有我的兩個女兒，總是能給我帶來無盡的快樂，還要感謝深蘭科技的創始人陳海波先生和首席戰略官王博士，兩位幫助我完成這本書，並提出了很多意見。

另外，感謝博文視點給我機會出版這本書，希望透過這本書能結識更多的開發者。還要感謝南海寶編輯在本書撰寫和出版過程中給予的指導和鼓勵。

目錄

01 機器學習和 TensorFlow 簡述

02 建置開發環境

03 以行動端為基礎的機器學習的開發方式和流程

04 建置 TensorFlow Mobile

05 用 TensorFlow Mobile 建置機器學習應用

06 TensorFlow Lite 的架構

07 用 TensorFlow Lite 建置機器學習應用

08 行動端的機器學習開發

09 TensorFlow 的硬體加速

10 機器學習應用架構

11 以行動裝置為基礎的機器學習的未來

機器學習和
TensorFlow 簡述

1.1 機器學習和 TensorFlow 的歷史及發展現狀

1.1.1 人工智慧和機器學習

我們先來看一下人工智慧（Artificial Intelligence）、機器學習（Machine Learning）、深度學習（Deep Learning）的定義。

人工智慧（英文縮寫為 AI）也稱機器智慧，指由人製造出來的機器所表現出來的智慧。

下面是機器學習的英文定義：

Machine learning is a core, transformative way by which we're rethinking how we're doing everything.

其中文含義是：機器學習是一種核心的、變革性的方式，它正在改變我們思考的方式。人工智慧（AI）是使事物變得聰明的科學，機器學習是一種開發人工智慧的技術。

深度學習（Deep Learning）是機器學習的分支，是一種以類神經網路為架構，對資料進行表徵學習的演算法。人工智慧的分類如圖 1-1 所示。

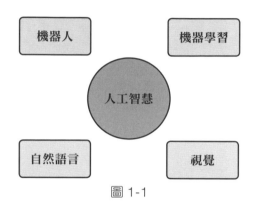

圖 1-1

人工智慧按產業分類大致可以分為下面幾種：機器學習、自然語言處理、機器人技術和視覺等。在機器學習裡深度學習是最近興起也是比較熱門的研究方面。自然語言處理和視覺的技術發展近幾年來越來越成熟，有的技術已被大規模應用。

深度學習帶來機器學習的革命。我們看到「深度學習」這個詞在搜索中的熱度近年來在快速攀升。arXiv 上的機器學習論文數量也在急遽增長。

深度學習（Deep Learning）是一種模仿人腦結構的機器學習。神經元是專注於某個特定方向的刺激（例如影像中物件的形狀、顏色和透明度），透過將多個神經元分層組合在一起而完成模擬人腦的方法。分層可以模擬大腦運算，隨著層數的增加，計算的功率和時間也會增加，進而加強計算的準確性。

下面來看一個圖片分類的實例:列出一張圖片,讓機器識別這張圖是一隻貓還是一條狗。這個機器由多層的神經網路結構組成,該結構中有很多參數,經過大量訓練之後,機器能識別出這張圖是一隻貓。如圖 1-2(圖片來源 https://becominghuman.ai/building-an-image-classifier-using-deep-learning-in-python-totally-from-a-beginners-perspective-be8dbaf22dd8)所示,展示了這個深度學習的過程。

深度學習並不是全新的事物,但為什麼在最近幾年有了極大突破?其中一個重要的原因是,人類發明了一個以深度神經網路為基礎的解決方案。

在 20 世紀五六十年代,神經網路的研究就已經出現了,在馬文明斯基和西摩帕爾特(1969)發表了一項關於機器學習的研究以後,神經網路的研究就停滯不前了。

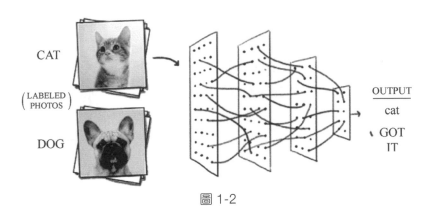

圖 1-2

他們發現了神經網路的兩個關鍵問題點。一個問題是,基本感知機無法處理互斥迴路;另一個問題是,電腦沒有足夠的能力來處理大型神經網路所需要的計算時間。在計算機具有更強的運算能力之前,神經網路的研究進展緩慢。但是隨著運算能力的增加,深度學習解決問題的精度已經超過其他機器學習方法。

以圖片識別為例，2011 年，機器識別的錯誤率是 26%，而人工識別的錯誤率只有 5%，所以這個時候的機器識別離實用有非常大的距離。到 2016 年，機器識別的錯誤率已經減少到 3% 左右，深度學習在該領域呈現出非常驚人的能力，這也是深度學習在影像識別領域吸引產業界大量關注的原因。

組成機器學習的三大要素是：資料、計算力和演算法（Data、Computation and Algorithm）。

1.1.2 TensorFlow

接下來看一下最近非常流行的，也是本書主要說明的機器學習架構 TensorFlow。

1 TensorFlow 的起源和發展歷史

TensorFlow 是一個開放原始碼軟體函數庫，用於完成各種感知和語言了解工作的機器學習。TensorFlow 被 50 個團隊用於研究和開發許多 Google 商業產品，如語音辨識、Gmail、Google 相簿和搜索，其中許多產品曾使用過其前任軟體 DistBelief。TensorFlow 最初由 Google 大腦團隊開發，用於 Google 的研究和產品開發，於 2015 年 11 月 9 日在 Apache 2.0 開放原始碼許可下發佈。

2010 年，Google 大腦建立 DistBelief 作為第一代專有機器學習系統。Google 的 50 個團隊在 Google 和其他 Alphabet 公司的商業產品中部署了 DistBelief 的深度學習神經網路，包含 Google 搜索、Google 語音搜索、廣告、Google 相簿、Google 地圖、Google 街景、Google 翻譯和 YouTube。

Google 安排電腦科學家如 Geoffrey Hinton 和 Jeff Dean，簡化和重構了 DistBelief 的程式庫，使其變成一個更快、更穩固的應用級程式庫，形成了 TensorFlow。

2009 年，Hinton 主管的研究團隊透過在廣義反向傳播方面的科學突破，相當大地加強了神經網路的準確性，使得神經網路的生成成為可能。值得注意的是，這個科學突破使得 Google 語音辨識軟體中的錯誤數減少了至少 25%。如圖 1-3（水平座標為年份，垂直座標為 Google 搜索的人工智慧和機器學習關鍵字數量）所示，2013 年以後，人工智慧和機器學習的關鍵字搜索數量有了相當大增長。

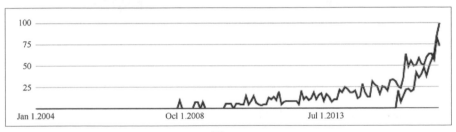

圖 1-3

TensorFlow 是 Google 大腦的第二代機器學習系統。從 0.8.0 版本（發佈於 2016 年 4 月）開始支援本機的分散式執行。從 0.9.0 版本（發佈於 2016 年 6 月）開始支援 iOS。從 0.12.0 版本（發佈於 2016 年 12 月）開始支援 Windows 系統。該移植程式主要是由微軟貢獻的。

TensorFlow 1.0.0 發佈於 2017 年 2 月 11 日。雖然推理的實現執行在單台裝置上，但 TensorFlow 也可以執行在多個 CPU 和 GPU（包含可選的 CUDA 擴充和圖形處理器通用計算的 SYCL 擴充）上。TensorFlow 可用於 64 位元的 Linux、macOS、Windows，以及行動計算平台（Android 和 iOS）。

TensorFlow 的計算使用有狀態的資料流程圖來表示。TensorFlow 的名字來自這種神經網路對多維陣列執行的操作,這些多維陣列被稱為「張量」。2016 年 6 月,Jeff Dean 稱:「在 GitHub 上有 1500 個函數庫提到了 TensorFlow,其中只有 5 個來自 Google。」

1.12.0 版本發佈於 2018 年 10 月,TensorFlow 2.0 的預覽版在 2019 年 3 月的 TensorFlow 開發者大會上發佈。

2 TensorFlow 的主要特性

TensorFlow 開放原始碼以來已有 500 多個 Contributor 及 11000 多個 Commit。利用 TensorFlow 平台在產品開發環境下進行深度學習的公司有 ARM、Google、UBER、DeepMind、京東等。Google 把 TensorFlow 應用到很多內部專案,如 Google 語音辨識、Gmail 電子郵件、Google 圖片搜索等。

TensorFlow 有以下幾個主要特性。

- 使用靈活:TensorFlow 是一個靈活的神經網路平台,採用圖型計算模型,支援 High-Level 的 API,支援 Python、C++、Go、Java 介面。

- 跨平台:TensorFlow 支援 CPU 和 GPU 的運算,支援桌上型電腦、伺服器、行動平台的計算,並從 0.12 版本開始支援 Windows 平台。

- 產品化:TensorFlow 支援從研究團隊發佈模型到產品開發團隊驗證模型的全過程,建置起模型研究到產品開發實作的橋樑。

- 高性能:在 TensorFlow 中採用了多執行緒、佇列技術及分散式訓練模型,可在多 CPU、多 GPU 的環境下對模型進行分散式訓練。

1.1.3 TensorFlow Mobile

其實 TensorFlow Mobile 並不是一個正式的名稱，它只是 TensorFlow 對
行動端裝置和 IoT 裝置的支援的總稱，它可以直接移植到 Android、iOS
和樹莓派等系統上。

1.1.4 TensorFlow Lite

2017 年 5 月 Google 宣佈從 Android Oreo（API level 26）開始，提供一
個專用於 Android 開發的軟體堆疊 TensorFlow Lite。按照官方的定義，
TensorFlow Lite 是為行動裝置和嵌入式裝置設計的機器學習軟體架構，
也是在行動和嵌入式裝置上執行機器學習模型的官方解決方案。它支援
在 Android、iOS 和其他作業系統上的低延遲和在較小二進位檔案裝置上
的機器學習推理。

TensorFlow Lite 在 2017 年 10 月發佈了第一個 Preview 版本，在官方網
頁 https://developers. googleblog.com/2017/11/announcing-tensorflow-lite.
html 上有以下說明：

*Lightweight Enables inference of on-device machine learning models with
a small binary size and fast initialization/startup.*

*Cross-platform A runtime designed to run on many different platforms,
starting with Android and iOS.*

*Fast Optimized for mobile devices, including dramatically improved model
loading times, and supporting hardware acceleration.*

透過該段説明，我們可以將 TensorFlow Lite 的特性歸納為以下三點：

（1）輕量級：使用小的二進位檔案和快速初始化（啟動），可以在裝置端訓練機器學習模型。

（2）跨平台：可以在 Android 和 iOS 的許多不同平台上執行。

（3）快速：針對行動裝置進行最佳化，包含顯著改進的模型載入時間和支援硬體加速。

如圖 1-4（圖片來源 https://techcrunch.com/2017/05/17/googles-tensorflow-lite-brings-machine- learning-to-android-devices）所示是 TensorFlow Lite 在 Google I/O 發佈時的情景。

圖 1-4

1.2 在行動裝置上執行機器學習的應用

為什麼要在行動端和 IoT 裝置上進行機器學習？我們先來討論下面幾個問題：

- 為什麼要在行動端上進行機器學習？

- 行動端上的機器學習要解決什麼問題？

- 行動端機器學習面臨的挑戰是什麼？

1.2.1 生態和現狀

為什麼要在行動裝置和嵌入式裝置上進行機器學習呢？在中國大陸網際網路的發展過程中，PC 網際網路已經日趨飽和，行動網際網路卻呈現爆炸式發展。中國網際網路資訊中心發佈的 2018 年網際網路發展報告資料顯示，截至 2018 年 6 月，中國手機網民超過 8.2 億，佔網民總數的 98.35%。隨著行動終端價格的下降及 WiFi 的廣泛使用，行動網民的數量呈現爆發趨勢。

被稱為「網際網路女皇」的瑪麗米柯爾在《2018 年網際網路趨勢報告》裡說，中國的行動網際網路企業在 2018 年迎來新的增長，中國網民人數已經超過 7.53 億，佔總人口的一半以上。行動資料流量消費相較去年上漲了 162%。

在智慧型手機市場，Android 和 iOS 是百分比最大的兩種作業系統，在全球已經啟動的 31 億智慧型手機中的百分比超過 95%，而 Android 在兩者之中又佔據了絕對優勢，截至 2017 年 11 月，Android 的市佔率為 75.9%，智慧型手機總計 23 億部。中國和印度是全球最大的兩個 Android 智慧型手機市場，百分比接近一半。而根據 Newzoo 發佈的《全球手機市場報告》，2018 年將再有 3 億部新手機被啟動，Android 手機的優勢將進一步擴大。

Android 作為作業系統，它的生態包含各種類型的裝置和儀器，從汽車、穿戴裝置、VR/AR 裝置到 IoT 裝置，組成了一個龐大的系統，如圖 1-5（圖片來源 https://www.itproportal.com/ features/iot-what-businesses-need-to-know/）所示。

圖 1-5

在世界物聯網博覽會發佈的《2017—2018 年中國物聯網發展年度報告》中顯示，2017 年全球物聯網裝置數量強勁增長，達到 84 億台，第一次超過人口數量。全球物聯網市場有望在十年內實現大規模普及，到 2025 年市場規模或將增長至 3.9MB~11.1MB 美金。

物聯網發展呈現一些新的特點和趨勢：一是全球物聯網裝置數量爆發式增長，物聯網解決方案逐漸成熟；二是中國大陸物聯網市場規模突破兆元,物聯網雲端平台成為競爭核心領域;三是物聯網細分領域熱度出現分化,技術演進驅動應用產品向智慧、便捷、低耗電方向發展。在 IoT 上執行的作業系統和應用,都要適用於物聯網的特性。

1.2.2 從行動優先到人工智慧優先

2017 年,Google 決定將公司戰略從行動優先轉變為人工智慧優先。雖然 Google 近十年來主要以行動優先,但很明顯調整表示公司看到了人工智慧和機器學習技術的極大潛力。2019 年,Google 花了很大力氣進行這種改變,並獲得了顯著的成果。

這裡要說一下 Google 的 CEO Sundar Pichai,這位可算作矽谷最成功的印度裔的高管,一開始加入 Google 的 Chrome 團隊,擔任 Chrome 的產品經理。Chrome 是一款非常成功的產品,現在佔據了瀏覽器市場超過 65% 的百分比,也是很多使用者包含筆者自己的預設瀏覽器。

當然,ChromeOS 並不能算世界級的產品。雖然 Chrome 在 Google 內和其他世界級的產品相比並不是特別成功,但是它在教育領域及針對政府和企業的專案中或許會有令人矚目的進展。有意思的是,很多一開始從事 ChromeOS 開發的工程師,後來開始 GoogleIoT 的研究,現在又在做新的作業系統的研究。

另外,Google 在經歷了幾年行動優先之後,在 2017 年,Sundar Pichai 正式在 GoogleI/O 上提出了「從行動優先到人工智慧優先」的戰略。這是在繼 ChormeOS 與 Android 整合後的非常重要的決定。我們可以看到,這幾年人工智慧已經成為一個非常熱門的投資領域。

1.2.3 人工智慧的發展

Jeff Dean 在人工智慧發展剛剛取得突破性進展的時候就意識到，Google 可能無法提供足夠的計算力來支援人工智慧的發展。後來，Google 啟動了 TPU 專案，透過硬體加速為人工智慧提供強有力的計算基礎。這表現在雲端運算和資料中心的 TPU POD 叢集的部署中，也表現在對 Edge TPU 的研發和對行動端及 IoT 裝置的進一步支援上。Google 很重視基礎技術和基礎專案的研究，並願意長時間地進行研發。後面，我們會介紹硬體加速的進展。如圖 1-6（圖片來源 https://www. datacenterknowledge. com/machine-learning/you-can-now-rent-entire-ai-supercomputer-google-cloud）所示是 Google 配備 TPU 的資料中心的照片。

圖 1-6

1.2.4 在行動裝置上進行機器學習的困難和挑戰

在行動裝置上進行機器學習是非常困難和具有挑戰性的，主要表現在以下幾個方面：

■ 行動裝置的 CPU 功率和電池電量非常有限。

■ 裝置和雲之間的連接非常有限。

■ 避免裝置和雲之間的大數據交換。

■ 解決裝置和雲之間的遺留問題。

■ 保護使用者的隱私資料。

隨著行動裝置變得越來越強大，我們將看到更多執行在行動裝置上的機器學習應用程式。

邊緣運算是一個新技術。Google 在 2016 年發佈 TPU（張量處理單元），並在其資料中心大量使用 TPU。Google 將 TPU 在資料中心的使用範圍擴充到不同的域，將來我們或許可以執行完全支援 AI 的行動裝置。

1.2.5 TPU

機器學習的發展一方面依賴軟體和演算法的加強，另一方面也離不開硬體的進步。2016 年 5 月，Google 發佈 TPU，一個專為機器學習和 TensorFlow 訂製的 ASIC。TPU 是一個可程式化的 AI 加速器，提供高傳輸量的低精度計算（如 8 位），用於使用或執行模型而非訓練模型。Google 在資料中心執行 TPU 長達一年多，發現 TPU 對機器學習提供一個數量級更優的每瓦特效能。

2017 年 5 月，Google 發佈第二代 TPU，此款 TPU 在 Google 的 Compute Engine 中是可用的。第二代 TPU 提供最高 180 teraflops 的效能，在組裝成 64 個 TPU 的叢集時提供最高 11.5 petaflops 的效能。如圖 1-7（圖片 來 源 https://blog.hackster.io/announcing-the-new-aiy-edge-tpu-boards-98f510231591）和圖 1-8（圖片來源 http://hi.sevahi.com/google-unveils-tiny-new-ai-chips-for-on- device-machine-learning/）所示分別是 Google 對外公佈的 TPU 和 Edge TPU 的圖片。

圖 1-7   圖 1-8

1.3　機器學習架構

本節將介紹 TensorFlow Mobile、TensorFlow Lite 等工業界現有的機器學習架構，以及其他業界廣泛使用的機器學習架構。另外，本節還將介紹對應行動裝置和 IoT 裝置的開發現狀。

TensorFlow 具有不同的建置方式，並從一開始就支援不同的平台，例如 Android、iOS 和樹莓派。值得一提的是，Google 的物聯網團隊早期就是在樹莓派上建置及執行 TensorFlow 的。

在本書中，我們主要討論和描述如何在 Android 裝置上建置 TensorFlow。

TensorFlow 支援 TensorFlow Mobile 和 TensorFlow Lite 兩個平台。本節將詳細討論兩者的使用方法和區別，以及未來的發展趨勢。

TensorFlow Mobile 有更好的支援全張量流的函數，我們可以在行動裝置上執行張量流，但是 TensorFlow Mobile 並沒有針對行動裝置進行高度最佳化。

TensorFlow Lite 針對行動裝置進行了高度最佳化，但作為一個新架構，它只具有有限的支援張量流功能。

1.3.1 CAFFE2

CAFFE（Convolutional Architecture for Fast Feature Embedding） 是 一個開放原始碼的深度學習架構，最初是在加州大學柏克萊分校開發的。CAFFE 是用 C++ 撰寫的，支援 Python 介面。

賈清揚博士在加州大學柏克萊分校攻讀博士學位期間建立了 CAFFE 專案。現在該專案有很多貢獻者，並在 GitHub 上進行託管。賈清揚博士現在是 Facebook 的人工智慧工程總監，CAFFE 的服務架構也因此多多少少帶上了 Facebook 的印記。

2017 年 4 月，Facebook 宣佈了 CAFFE2，其中包含 Recurrent Neural Networks 等新功能。2018 年 3 月底，CAFFE2 被合併到 PyTorch。

CAFFE2 有不短的歷史，對行動和嵌入式裝置的支援也比較好。所以，很多在行動端進行機器學習開發的人會首選 CAFFE2。

1.3.2 Android NNAPI

Android NNAPI（Neural Networks API）是一個 Android C API，專門為在行動裝置上對機器學習進行密集型運算而設計。NNAPI 主要在為建置和訓練神經網路的更進階機器學習架構（如 TensorFlow Lite、CAFFE2或其他）提供一個基礎的功能層。API 適用於執行 Android 8.1（API 等級為 27）或更新版本的所有裝置。

1.3.3 CoreML

在蘋果發佈的 iOS11 中，有一個新的軟體架構叫作 CoreML。使用CoreML，可以將經過訓練的機器學習模型整合到應用中。CoreML 是自然語言處理專業領域的架構和功能的基礎。CoreML 支援用於影像分析的 Vision、用於自然語言處理的 Foundation，以及用於評估學習決策樹的 GameplayKit。

CoreML 本身建立在低階基本操作之上，如 Accelerate、BNNS 及 MetalPerformance Shaders。CoreML 針對元件內的效能進行了最佳化，可大幅地減少記憶體佔用和耗電。在裝置上執行可確保使用者資料的隱私性，並確保在網路連接不可用時，應用程式仍可正常執行並做出回應。想要了解更多的資訊可以在蘋果的開發者網頁上進行檢視。

以上蘋果的這些描述是非常有意思的，這裡提到了機器學習的特定技術，例如視覺、自然語音等，但是沒有特別提到機器學習或人工智慧，是一段比較嚴謹的描述。

隨著人工智慧和機器學習的發展，現在可選擇的機器學習架構越來越多。例如由亞馬遜、微軟和英特爾共同開發現已成為 Apache 開放原始碼

專案的 MXNet，由 NVIDIA 開發的可以實現硬體加速的 TensorRT，由
阿里巴巴開放原始碼的深度學習架構 X-Deep Learning 和機器學習平台
PAI，以及由百度開放原始碼的深度學習架構 PaddlePaddlc。這還不包含
一些具有更長歷史的機器學習架構，以及一些原生的計算架構。

作為普通開發者，在選擇上會有一些難度。但是，由於機器學習和人工
智慧的複雜性，我們也看到了一些融合的趨勢。舉例來說，新的學習架
構和舊的學習架構的無縫融合和整合，學習架構對於 GPU、硬體和高計
算的支援。開發者在選擇架構的時候要考慮哪個架構適合自己，同時也
要考慮這個架構的未來發展性。在 Wiki 裡，有一個關於各種架構的比較
分析，如圖 1-9 所示。

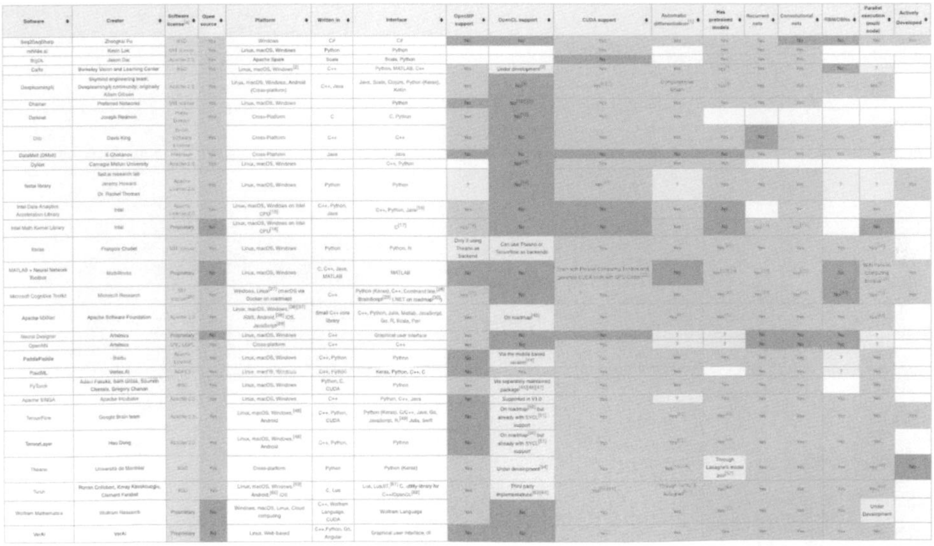

圖 1-9 機器學習架構的比較

註：該圖比較大，讀者可到 https://en.wikipedia.org/wiki/Comparison_of_deep_learning_
software 上查閱。

1.3.4 樹莓派（Raspberry Pi）

依據樹莓派網站 https://www.raspberrypi.org/products/raspberry-pi-3-model-b-plus 的介紹，它們的最新產品是 Raspberry Pi 3 Model B+。它設定有一個 4 核心 64bit 的 Cortex-A53 處理器，主頻 1.4GHz。同時也支援無線網路、藍牙等。秉承一貫的風格，樹莓派的開發板小巧同時又能提供相當的運算能力。樹莓派被用作很多 IoT 或嵌入式開發的參照板。機器學習需要在這種運算能力、儲存能力等都有限的平台上表現出優越的效能，才能在更廣闊的範圍內獲得推廣和應用。

Chapter

02

建置開發環境

這一章，我們將學習怎樣建置 TensorFlow 的開發環境。在介紹建置 TensorFlow 特有的開發環境之前，我們還要介紹一些同類軟體開發環境的建置方法。

2.1 開發主機和裝置的選擇

許多開發人員使用 Windows 作為開發平台，但 Android 不支援 Windows 作為開發平台。多數 Google 工程師使用訂製的 Linux 作為主要平台，因為使用相同的平台可以避免任何可能由平台引起的角落案例問題。建議開發者安裝 Ubuntu 16 或使用 macOS。熟悉 Linux 對於 Android 開發非常有幫助，因為 Linux 上的許多概念和指令工具都可以在 Android 上使用。

Android 開發人員可以在 AOSP（Android Open-Source Project，Android 開放原始程式碼專案）中開發應用程式，但這種開發方式非常複雜。因此，大多數開發人員選擇 Android Studio 作為開發平台。

在此，筆者鼓勵開發人員使用以 AOSP 開發標準為基礎的獨立應用程式。很多 OEM 和 ODM 都是從 AOSP 開始開發訂製的 Android 的，但是，隨著開發規模和程式規模的不斷增長，開發人員應該考慮將平台開發和應用程式開發分開。透過這種方式，應用程式開發將只依賴於平台 SDK 而非原始程式碼，並且應用程式可以做更快的開發反覆運算、測試和部署。

2.2 在網路代理環境下開發

在網路代理或防火牆後面進行開發，都是非常困難的。因為，很多公司仍然使用 Windows 作為主要的開發平台，而且很多公司的 IT 部門對開放原始碼社群沒有很好的支援。我們要逐一解決這些問題。對於代理人背後的開發者，開發人員必須正確設定 GitHub.com 等網站。

在網路代理環境下開發，會有很多事情需要處理。首先，需要正確設定代理、使用者和密碼，這種設定可能會分散到多個檔案中，因此請嘗試記下更改的內容及執行此操作的最佳方法。

接下來解決認證問題，即許多網站的識別使用者功能。簡單的方法可能是設定設定跳過或忽略身份驗證，但大多數工具都有自己的設定，所以必須將它們分開設定。有些工具在內部會呼叫其他工具，在設定過程中需要確定哪個工具會出現問題。在開發過程中最壞的情況是：需要撰寫程式，重新建置工具或在工具中增加代理支援。為透過代理從 Maven 下載，筆者透過重建 Git 來支援 OpenSSL 和 Hack Bazel。

在 Ubuntu 16.04 上，筆者對以下工具參數進行了設定：

- .bashrc
- apt
- CNTLM
- git
- curl
- wget
- Bazel

實際設定參數和內容如表 2-1 所示。

表 2-1　實際設定參數和內容

設定參數	內容	設定參數	內容
.bashrc	proxy	curl	authentication
apt	proxy	wget	authentication
CNTLM	authentication	Bazel	maven download
git	authentication		

2.3　整合式開發環境 IDE

2.3.1 Android Studio

Android Studio 是 Google 支援的官方 IDE。

下載並安裝 Android Studio。Android Studio 支援 Gradle，但在 2018 年 6 月以前，3.1.2 版本不支援 Bazel 0.13。希望 Google Android Studio 團隊和 Blaze 團隊能夠更進一步地同步並提供無縫支援，否則 Android 社群很難採用 Bazel。

2.3.2 Visual Studio Code

Visual Studio Code 是由微軟發佈的開發整合環境。其官方網站上寫道：

Visual Studio Code 是一個輕量級但功能強大的原始程式碼編輯器，可在桌面上執行，可用於 Windows、macOS 和 Linux。它支援 JavaScript、TypeScript 和 Node.js 等指令稿，並且具有豐富的語言（如 C ++、C #、Java、Python、PHP、Go）和執行時期（如 .NET 和 Unity）擴充生態系統。

Visual Studio Code 的安裝步驟非常簡單。安裝完畢後，根據需要繼續安裝 Bazel 擴充和其他語言的擴充。如圖 2-1 所示為 Bazel 的擴充頁面。

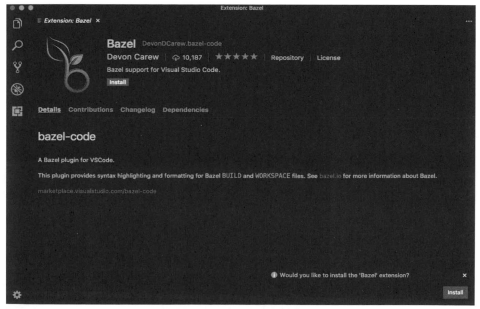

圖 2-1　Bazel 的擴充頁面

開發人員也可以安裝一些 Android 外掛程式，圖 2-2 展示了安裝 Bazel 外掛程式的相關資訊。

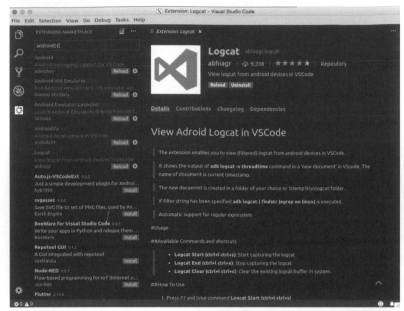

圖 2-2 安裝 Bazel 外掛程式

安裝完相關外掛程式後，Visual Studio Code 應該正常執行，包含 C++/
Java/Python 等語言支援功能和自動完成功能，圖 2-3 為 Visual Studio
Code 的編輯介面。

圖 2-3 Visual Studio Code 的編輯介面

不得不説,微軟在支援開放原始碼社群方面做得確實很好,這引起了很多開發者的關注。至少,筆者在寫這本書的時候,使用的 IDE 就是 Visual Studio Code。

2.3.3 其他 IDE

除 Visual Studio Code 外,還有 IntelliJ、Eclipse 等多種工具供開發者選擇使用,開發者可根據個人習慣選擇不同的工具。在過去很長一段時間內,Eclipse、Emacs 和 Vi 佔據了 IDE 的主流市場,隨後過渡到 IntelliJ,目前以 Android Studio 為主。還有另一個強大的以雲端為基礎的 IDE 在 Google 中越來越受歡迎。

2.4 建置工具 Bazel

大多數開發人員使用 Android Studio + Gradle + Maven 作為他們的日常工具,這種設定已經足夠了。不過,筆者強烈建議使用 Bazel,它是 Google 內部工具的開放原始碼版本。Bazel 具有很多重要的功能,最新版本是 Bazel 0.22,與 1.0 正式發佈版本非常接近。很多離開 Google 的工程師,仍會繼續使用 Bazel。由 Facebook 開發的 Buck 與 Bazel 的功能十分類似。許多初創公司使用的也是 Bazel 工具。

Bazel 的建置方法很簡單,在 Linux 上執行下面的指令就可以安裝 Bazel:

```
sudo apt-get install bazel
sudo apt-get upgrade bazel
```

Android Studio 也支援 Bazel。在 Android Studio 上安裝 Bazel 外掛程式即可使用，安裝 Bazel 外掛程式的步驟如圖 2-4 所示。

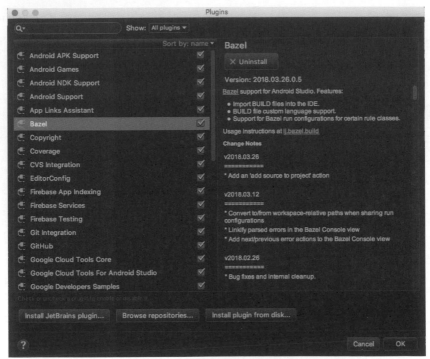

圖 2-4 在 Android Studio 上安裝 Bazel 外掛程式

TensorFlow 中支援 Gradle、Make 和 iOS 的建置，Google 內部不支援 Gradle，只是在開放原始碼的時候會測試。

2.4.1 Bazel 產生偵錯

產生偵錯指令為：

```
bazel build --compilation_mode=dbg //tensorflow/lite/toco
```

2.4.2 Bazel Query 指令

如果讀者想找出 tensorflow/python/saved_model 和 tensorflow/core/kernels: slice_op 運算元之間的關係，可以執行 bazel query "somepath (tensorflow/ python/saved_model, tensorflow/ core/kernels:slice_op)"，執行結果如下：

```
//tensorflow/python/saved_model:saved_model
//tensorflow/python/saved_model:builder
//tensorflow/python:lib
//tensorflow/python:pywrap_tensorflow
//tensorflow/python:pywrap_tensorflow_internal
//tensorflow/python:_pywrap_tensorflow_internal.so
//tensorflow/python:tf_session_helper
//tensorflow/core:all_kernels
//tensorflow/core/kernels:array
//tensorflow/core/kernels:slice_op
```

如果讀者想找出 //tensorflow/contrib/lite/java/demo/app/src/main:TfLiteCamera Demo 和 //tensorflow/contrib/lite/kernels:builtin_ops 運算元之間的相依關係，可以執行 bazel query "somepath(//tensorflow/contrib/lite/java/demo/app/ src/main: TfLiteCameraDemo,//tensorflow/contrib/lite/kernels:builtin_ops)"，執行結果如下：

```
//tensorflow/lite/java/demo/app/src/main:TfLiteCameraDemo
//tensorflow/lite/java:tensorflowlite
//tensorflow/lite/java:tflite_runtime
//tensorflow/lite/java:libtensorflowlite_jni.so
//tensorflow/lite/java/src/main/native:native
//tensorflow/lite/kernels:builtin_ops
```

除 Android 特有的設定外，我們還要做 TensorFlow 的標準設定，實際的步驟請參考網頁 https://docs.bazel.build/versions/master/user-manual.html。

2.5 載入 TensorFlow

TensorFlow 的安裝比較簡單，官網上提供了詳細的說明，實際內容請參閱 TensorFlow 的網上連結。

需要提醒的是，建議使用 Virtualenv 來安裝 TensorFlow，安裝完 TensorFlow 和 GPU（圖形加速器）支援後，需要驗證。

由於某些使用者在安裝 TensorFlow GPU 支援時會遇到問題，因此接下來將介紹如何安裝 GPU 支援。

首先，開發者要在 Developer Nvidia Website 上註冊。

然後，按照此連結安裝 GPU。

接著，還需要安裝 CUDA Toolkit 9.0，tensorflow.org 中的連結始終指向最新的 CUDA 版本，現在是 9.2 版本。但是不要使用 9.2 版本，除非 TensorFlow 支援它。請使用上面連結的 CUDA 9.0 版本。

同樣，請下載並安裝 cuDNN v7.1.4 for CUDA 9.0，tensorflow.org 中的連結指向的最新版 cuDNN 是 CUDA 9.2 的 v7.1.4 版本。安裝並執行以下指令：

```
$ nvcc -V
nvcc: NVIDIA (R) Cuda compiler driver
Copyright (c) 2005-2017 NVIDIA Corporation
Built on Fri_Sep__1_21:08:03_CDT_2017
Cuda compilation tools, release 9.0, V9.0.176
```

接著，執行指令 "$ nvidia-smi"，獲得以下結果：

```
Fri Jun 1522:21:082018
+-----------------------------------------------------------------------------+
```

```
| NVIDIA-SMI 384.130                    Driver Version: 384.130           |
|-------------------------------+----------------------+------------------------+
| GPU  Name        Persistence-M| Bus-Id        Disp.A | Volatile Uncorr. ECC |
| Fan  Temp  Perf  Pwr:Usage/Cap|         Memory-Usage | GPU-Util  Compute M. |
|===============================+======================+======================|
|   0  Quadro K600          Off | 00000000:05:00.0 Off |                  N/A |
| 25%  48C    P0    N/A / N/A |     0MiB /   979MiB |      0%      Default |
+-------------------------------+----------------------+------------------------+

+-------------------------------------------------------------------------+
| Processes:                                                  GPU Memory |
|  GPU       PID   Type   Process name                        Usage      |
|=========================================================================|
|  No running processes found                                            |
+-------------------------------------------------------------------------+
```

此外，還要在 CUDA 範例程式中執行 deviceQuery，以確保 GPU 正常執行。

```
Device 0: "Quadro 600"
CUDA Driver Version / Runtime Version          9.0 / 9.0
CUDA Capability Major/Minor version number:    2.1
Total amount of global memory:                 962 MBytes (1009254400 bytes)
```

執行結果如下：

```
Quadro M6000
```

如果你看到類似的結果，説明你的顯示卡可以支援 TensorFlow。

然後，我們執行下面的指令：

```
$ ./bin/x86_64/linux/release/deviceQuery
```

執行結果會顯示顯示卡的版本編號和各種效能資料：

```
./bin/x86_64/linux/release/deviceQuery Starting...

 CUDA Device Query (Runtime API) version (CUDART static linking)

Detected 1 CUDA Capable device(s)

Device 0: "Quadro M600024GB"
  CUDA Driver Version / Runtime Version          9.0 / 9.0
  CUDA Capability Major/Minor version number:  5.2
  Total amount of global memory:               24467 MBytes (25655836672 bytes)
  (24) Multiprocessors, (128) CUDA Cores/MP:   3072 CUDA Cores
  GPU Max Clock rate:                          1114 MHz (1.11 GHz)
  Memory Clock rate:                           3305 Mhz
  Memory Bus Width:                            384-bit
  L2 Cache Size:                               3145728 bytes
  Maximum Texture Dimension Size (x,y,z)       1D=(65536), 2D=(65536,
65536), 3D=(4096, 4096, 4096)
  Maximum Layered 1D Texture Size, (num) layers  1D=(16384), 2048 layers
  Maximum Layered 2D Texture Size, (num) layers  2D=(16384, 16384), 2048
layers
  Total amount of constant memory:             65536 bytes
  Total amount of shared memory per block:     49152 bytes
  Total number of registers available per block: 65536
  Warp size:                                   32
  Maximum number of threads per multiprocessor: 2048
  Maximum number of threads per block:         1024
  Max dimension size of a thread block (x,y,z): (1024, 1024, 64)
  Max dimension size of a grid size    (x,y,z): (2147483647, 65535, 65535)
  Maximum memory pitch:                        2147483647 bytes
  Texture alignment:                           512 bytes
  Concurrent copy and kernel execution:        Yes with 2 copy engine(s)
  Run time limit on kernels:                   Yes
```

```
Integrated GPU sharing Host Memory:            No
Support host page-locked memory mapping:       Yes
Alignment requirement for Surfaces:            Yes
Device has ECC support:                        Disabled
Device supports Unified Addressing (UVA):      Yes
Supports Cooperative Kernel Launch:            No
Supports MultiDevice Co-op Kernel Launch:      No
Device PCI Domain ID / Bus ID / location ID:   0 / 4 / 0
Compute Mode:
   < Default (multiple host threads can use ::cudaSetDevice() with device
simultaneously) >

deviceQuery, CUDA Driver = CUDART, CUDA Driver Version = 9.0, CUDA Runtime
Version = 9.0, NumDevs = 1
Result = PASS
$ nvidia-smi
+-----------------------------------------------------------------------------+
| NVIDIA-SMI 384.130                 Driver Version: 384.130                   |
|-------------------------------+----------------------+----------------------+
| GPU  Name        Persistence-M| Bus-Id        Disp.A | Volatile Uncorr. ECC |
| Fan  Temp  Perf  Pwr:Usage/Cap|         Memory-Usage | GPU-Util  Compute M. |
|===============================+======================+======================|
|   0  Quadro M600024GB    Off  | 00000000:04:00.0 On  |                  Off |
| 25%   41C    P8    20W / 250W |    488MiB / 24467MiB |     0%       Default |
+-------------------------------+----------------------+----------------------+

+-----------------------------------------------------------------------------+
| Processes:                                                       GPU Memory |
|  GPU       PID   Type   Process name                             Usage      |
|=============================================================================|
|    0      2183      G   /usr/lib/xorg/Xorg                            319MiB |
|    0      3796      G   compiz                                         92MiB |
|    0      6095      G   ...-token=32ADD0D4261B4355966B2810A61BBF37     72MiB |
+-----------------------------------------------------------------------------+
```

最後，還要安裝 TensorFlow GPU：

```
(tensorflow)$ pip install --upgrade tensorflow       # for Python 2.7
(tensorflow)$ pip3 install --upgrade tensorflow      # for Python 3.n
(tensorflow)$ pip install --upgrade tensorflow-gpu  # for Python 2.7 and GPU
(tensorflow)$ pip3 install --upgrade tensorflow-gpu # for Python 3.n and GPU
```

安裝成功之後，可以用下面的指令確認：

```
(tensorflow) $ python
Python 2.7.12 (default, Dec  42017, 14:50:18)
[GCC 5.4.020160609] on linux2
Type "help", "copyright", "credits" or "license" for more information.
>>> import tensorflow as tf
>>> hello = tf.constant("hello")
>>> sess = tf.Session()
**2018-06-2006:54:34.284161: I tensorflow/core/platform/cpu_feature_ guard.
cc:140] Your CPU supports instructions that this TensorFlow binary was not
compiled to use: AVX2 FMA**
**2018-06-2006:54:34.460555: I tensorflow/core/common_runtime/gpu/gpu_
device.cc:1356] Found device 0 with properties: **
**name: Quadro M600024GB major: 5 minor: 2 memoryClockRate(GHz): 1.114**
**pciBusID: 0000:04:00.0**
**totalMemory: 23.89GiB freeMemory: 23.29GiB**
**2017-05-2006:54:34.460600: I tensorflow/core/common_runtime/gpu/gpu_
device.cc:1435] Adding visible gpu devices: 0**
**2017-05-2006:54:34.708584: I tensorflow/core/common_runtime/gpu/gpu_
device.cc:923] Device interconnect StreamExecutor with strength 1 edge
matrix:**
**2017-05-2006:54:34.708635: I tensorflow/core/common_runtime/gpu/gpu_
device.cc:929]      0 **
**2017-05-2006:54:34.708644: I tensorflow/core/common_runtime/gpu/gpu_
device.cc:942] 0:   N **
**2017-05-2006:54:34.709069: I tensorflow/core/common_runtime/gpu/gpu_
device.cc:1053] Created TensorFlow device (/job:localhost/replica:0/task:0/
```

```
device:GPU:0 with 22598 MB memory) -> physical GPU (device: 0, name**:
Quadro M600024GB, pci bus id: 0000:04:00.0, compute capability: 5.2)**
>>> print(sess.run(hello))
hello
```

上面的程式明確地顯示，開發者正在使用 GPU ！如果看不到這段程式，
說明開發者並沒有成功安裝 GPU，而是在使用 CPU。

2.6 文件

在本章的最後，筆者想分享一下文件的重要性。很多工程師不夠重視文
件，只有在軟體發佈時才會倉促寫一些文件，如果公司的管理層不重
視，文件就會流於空談。文件的作用是為了交流，便於團隊間的溝通學
習，用於未來的你和現在的你交流，所以文件要清楚、簡潔。

這裡筆者推薦一下 Markdown 檔案格式，比較權威的定義是，Markdown
是一種輕量級標記語言，創始人為約翰格魯伯（John Gruber）。它允許人
們「使用易讀易寫的純文字格式撰寫文件」。它非常輕，普通的文字編輯
器就可以使用，可隨時隨地寫作，不用額外購買其他軟體。另外，它的
語法非常簡單，常用的語法用 15 分鐘就可以記住。

Markdown 的表現形式也非常豐富，成文後的效果和 Word、PDF 沒有太
大的區別。這個工具非常好用，具有豐富的表現形式，而且可使用通用
管理工具進行管理，例如透過 git 就可以實現多人同時共用工作的目的。
本書就是用 Markdown 寫成的，可轉成 Word 或 PDF 文件，從第一次執
行 git check in 指令，大概經歷了 200 多次入倉，才算基本成書。

以行動端為基礎的機器學習的開發方式和流程

在介紹各種機器學習的開發之前，讓我們先大概介紹一下以行動裝置機器學習為基礎的開發方式和流程。考慮到行動裝置的特殊性，以行動裝置為基礎的機器學習與傳統的以網頁、雲端運算和資料中心為基礎的機器學習的方法有很大不同。以往的系統很多都是相對靜態和可控的，而且理想狀態下的運算資源是無限的，團隊可以更專注於演算法和系統的加強。

3.1 開發方式和流程簡介

現有的開發和應用大概可以分為以下四種。

第一種，在本機或雲端進行機器學習的訓練，訓練後的模型直接執行在行動裝置上。在這種情況下，雲端和裝置是完全分離的。在裝置上進行推理和運算。

第二種,和第一種相似,但是會以雲端或其他方式對在裝置上的機器學習進行不斷地更新和修正。在這種方式下,行動端也是以推理為主的。

第三種,即完全的雲端和行動端裝置互動的機器學習方式。雲端對正在裝置上進行的機器學習進行監測和修改,把修改過的模型發送到行動端,進行新的機器學習的推測。同時,裝置會把新的機器學習的結果發送到雲端。在雲端,做進一步地訓練,把新的增量結果發送到裝置上。這樣,雲端和裝置端就形成了一個完整的閉環。在這種方式下,基本還是延續了雲端作為機器學習訓練的主要資源,裝置端作為機器學習推理的主要資源。

第四種,較前三種更為複雜。它考慮到機器裝置數量眾多,所以同時進行機器學習的訓練和推理,並把機器學習訓練和推理的結果發送到雲端,在雲端進行進一步的訓練,並把訓練後的新的模型發送回裝置端。

第一種方式可能更適用於應用的原型的開發,或新模型開發的預研階段;第二、三種方式比較常見,已經大規模地應用在開發中;第四種比較複雜,需要雲端和裝置端的高度配合和比較複雜的演算法設計,這種應用比較少。圖 3-1 展示了幾種機器學習方式。

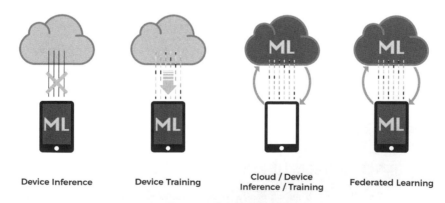

圖 3-1 機器學習方式

典型的以行動端為基礎的開發流程如下：

- 建置雲端或本機的模型和測試。

- 行動端的執行、測試和改進。

- 行動端產品化及使用者的使用。

由於技術資源和設計規模的限制，行動端的機器學習開發仍然嚴重依賴於行動裝置以外的開發。它與行動應用程式的開發十分類似，通常在桌面機上設計和建置應用程式，然後在行動裝置上執行應用程式。我們還要將來自行動裝置的資料與應用程式連接起來，並反覆運算回原始設計。然後，啟動另一個循環，以進一步改進應用程式。這個流程如圖 3-2 所示。

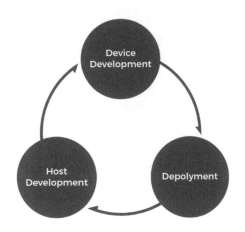

圖 3-2　行動端的機器學習開發流程

對於裝置上的機器學習，我們還要在主機裝置上開發原始模型，評估，並將其整合到行動應用程式中。我們從執行結果和記錄檔中收集資訊，並與期望值進行比較，再循環到下一個開發週期。

3.2 使用 TPU 進行訓練

現在大部分人工智慧訓練，都是在雲端或本機機上進行的。在本機機上用 GPU 進行訓練是比較普遍的方式，下面介紹 Google 用 TPU 進行機器學習訓練的方法。

首先開發者要註冊一個 Google 帳號，這個帳號不一定是 Gmail 的，帳號也可以用各個公司的郵寄地址。有了 Google 帳號以後，就可以在 Google 雲進行機器學習的訓練了。注意，Google 雲和 TPU 的使用現在都不是免費的。

首先，用 Google 帳號登入 Google 雲，新增一個專案，這裡新增了一個名為 tpuml 的專案，如圖 3-3 所示。

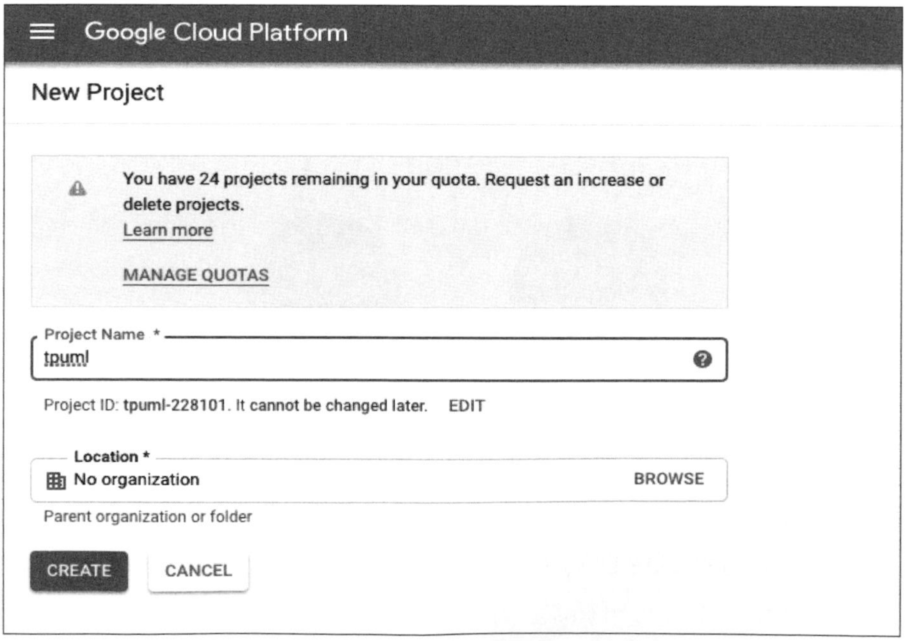

圖 3-3 專案頁面

然後，啟動機器學習 API 和 TPU 的 API。如圖 3-4 所示是啟動機器學習
API 的頁面；如圖 3-5 所示是啟動 TPU 的 API 頁面，可以看到使用時間
和費用。

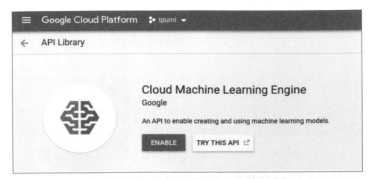

圖 3-4　啟動機器學習 API 的頁面

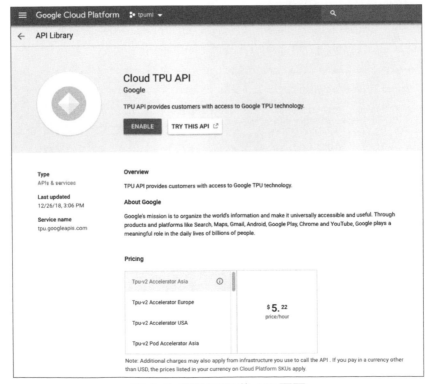

圖 3-5　啟動 TPU 的 API 頁面

Google 雲和機器學習服務不是免費的，需要啟動付費。首先，取得 TPU 的帳號，然後啟動 Shell，啟動收費頁面如圖 3-6 所示。

圖 3-6 啟動收費頁面

專案在建立時會產生一個虛擬機器，如圖 3-6 所示，藍色部分是虛擬機器相對應的 Shell。同時，每個專案又會分配一個 Docker 影像，如圖 3-6 所示，紅色的圖示可以啟動對應的 Shell。啟動 Shell 後，使用下面的指令，可以獲得 TPU 的帳號。

```
$ curl -H "Authorization: Bearer $(gcloud auth print-access-token)" https://
ml.googleapis.com/v1/projects/${PROJECT}:getConfig
/google/data/ro/teams/cloud-sdk/gcloud/google3/third_party/py/cryptography/
hazmat/primitives/constant_time.py:26: CryptographyDeprecationWarning:
Support for your Python version is deprecated. The next version of
cryptography will remove support. Please upgrade to a 2.7.x release that
supports hmac. compare_digest as soon as possible.
{
  "serviceAccount": "service-1049093715907@cloud-ml.google.com.iam.
gserviceaccount.com",
  "serviceAccountProject": "557869968812",
  "config": {
    "tpuServiceAccount": "service-557869968812@cloud-tpu.iam.
gserviceaccount.com"
  }
}
```

在取得帳號後，把它加到「收費帳戶」裡。如圖 3-7 所示是啟動收費帳號的頁面。這裡提醒一下讀者，如果你是個人開發者，要確認 TPU 的收費方式，確保收到的帳單是正確的。

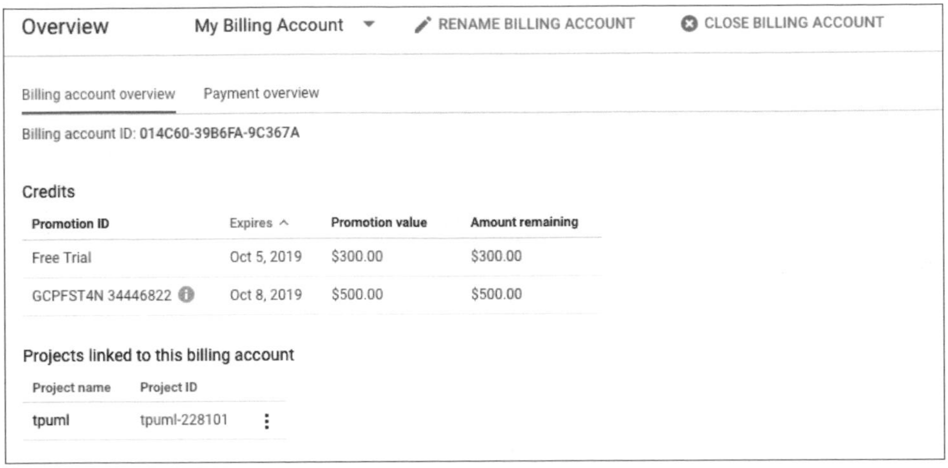

圖 3-7　啟動收費帳號的頁面

這時，就可以使用 TPU 進行機器學習的訓練了。

首先，列印目前 ctpu 的設定狀況，指令如下：

```
$ ctpu print-config
ctpu configuration:
        name: project_name
        project: tpuml-228101
        zone: us-central1-b
If you would like to change the configuration for a single command
invocation, please use the command line flags.
```

然後，執行 ctpu up 指令，啟動 TPU。

```
cloudshell:~ (tpuml-228221)$ ctpu up
ctpu will use the following configuration:
```

```
Name:                   project_name
Zone:                   us-central1-b
GCP Project:            tpuml-228221
TensorFlow Version:     1.12
VM:
    Machine Type:       n1-standard-2
    Disk Size:          250 GB
    Preemptible:        false
Cloud TPU:
    Size:               v2-8
    Preemptible:        false

OK to create your Cloud TPU resources with the above configuration? [Yn]: y
```

當 ctpu up 指令執行完畢後，驗證 Shell 提示符號已從 username@project
更改為 username@ tpuname，這表明目前已登入計算引擎 VM 中。

接下來，建一個叫作 tpuml-bucket 的儲存區域，如圖 3-8 所示。

圖 3-8　儲存頁面

在這裡，我們直接使用 TPU 虛擬機器附帶的程式進行訓練。

首先，設定 GCS_BUCKET_NAME：

```
export GCS_BUCKET_NAME=tpuml-bucket
```

然後，執行下面的指令，就可以進行 MNIST 的訓練。MNIST 是手寫數
字的資料集，使用這個資料集進行訓練，機器學習系統可以分辨出手寫
的數字。

```
$ python /usr/share/tensorflow/tensorflow/examples/how_tos/reading_data/
convert_to_records.py --directory=./data
$ gunzip ./data/*.gz
$ gsutil cp -r ./data gs://$GCS_BUCKET_NAME/mnist/data
$ python /usr/share/models/official/mnist/mnist_tpu.py --data_dir=gs:
//$GCS_BUCKET_NAME/mnist/data/ --model_dir=gs://$GCS_BUCKET_NAME/mnist/model
--tpu=$TPU_NAME
```

執行下面的指令，可以訓練 ResNet-50。

```
$ python \
    /usr/share/tpu/models/official/resnet/resnet_main.py \
    --data_dir=gs://cloud-tpu-test-datasets/fake_imagenet \
    --model_dir=gs://$GCS_BUCKET_NAME/resnet \
    --tpu=$TPU_NAME
```

ResNet-50 訓練的中間結果如圖 3-9 所示。

圖 3-9　ResNet-50 訓練的中間結果

在訓練的過程中，可以透過另外一個 Shell 啟動 TensorBoard：

```
$ tensorboard -logdir gs://$GCS_BUCKET_NAME/resnet &
```

TensorBoard 訓練的中間結果如圖 3-10 所示。

圖 3-10 TensorBoard 訓練的中間結果

TensorBoard 訓練的中間結果會被儲存在剛才建立的 Bucket 裡,如圖 3-11 所示顯示了儲存訓練的結果。

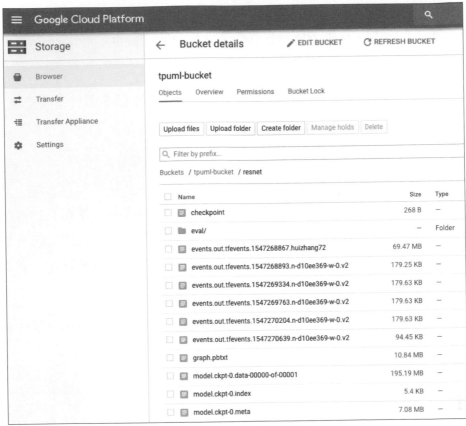

圖 3-11 儲存訓練的結果

3.3 裝置端進行機器學習訓練

前面我們學習了使用 Google 雲和 TPU 進行機器學習的訓練。Google 雲
提供了一個比較完整的機器學習開發環境,使用者體驗和本機機沒有什
麼差別。

下面我們將使用 Google 開放原始碼的物體識別模型，訓練出一個可以用在行動端的模型。模型的原始程式碼位址為 https://github.com/tensorflow/models/blob/master/research/object_detection。

我們繼續使用上面建立的 tpuml 專案，請留意下面三個名稱的不同意義：

```
Project name: tpuml
Project ID: tpuml-22801
Project number: 10909371907
```

project name 是專案名稱（可以不唯一），project ID 是程式中真正的專案名稱（唯一的），project number 是唯一的專案數字編號。我們以前提到了 Google 經常使用 Protobuf 作為資料定義的格式，估計在這裡，一個域（Field）是名稱，一個域是 ID，這也是常用的定義方式。

另外，為這次訓練，我們新增了一個叫作 tpuml-bucket/object 的儲存區域，並且設定對應的環境變數，程式如下：

```
$ export YOUR_GCS_BUCKET=tpuml-bucket/object
```

這次我們將要訓練一個能識別寵物的機器學習模型，選用 Oxford-IIIT Pets 資料集。首先，要把資料下載下來；然後，轉換成 TensorFlow 可以使用的資料格式 TFRecord；最後，上傳到剛才建立的雲端儲存裡。

接下來介紹 TFRecord。

TFRecord 是一種 TensorFlow 對應的檔案格式，它是透過 tf_record.py 指令稿（https:// github.com/ tensorflow/tensorflow/blob/master/tensorflow/python/lib/io/tf_record.py）來實現的。TFRecord 可以支援壓縮格式，是 TensorFlow 通用的資料儲存檔案格式。

由於接下來要用到 object_detection，所以我們要先下載原始程式碼：

```
$ git clone https://github.com/tensorflow/models.git
$ cd models/research/
```

然後，下載資料和處理資料：

```
# 下載資料集
$ wget http://www.robots.ox.ac.uk/~vgg/data/pets/data/images.tar.gz
$ wget http://www.robots.ox.ac.uk/~vgg/data/pets/data/annotations.tar.gz

# 解壓下載的資料集
$ tar -xvf images.tar.gz
$ tar -xvf annotations.tar.gz

# 在 models/research 裡
$ python object_detection/dataset_tools/create_pet_tf_record.py \
    --label_map_path=object_detection/data/pet_label_map.pbtxt \
    --data_dir=`pwd` \
    --output_dir=`pwd`

# 在 tensorflow/models/research 裡
$ gsutil cp pet_faces_train.record-* gs://${YOUR_GCS_BUCKET}/data/
$ gsutil cp pet_faces_val.record-* gs://${YOUR_GCS_BUCKET}/data/
$ gsutil cp object_detection/data/pet_label_map.pbtxt gs: //${YOUR_GCS_
BUCKET}/data/pet_label_map.pbtxt
```

下面，我們下載 COCO 預訓練模型進行遷移學習（Transfer Learning）。
如果從頭開始訓練物體探測器，那麼會需要大量的時間。為了加速訓
練，我們將採用在不同資料集（COCO）上訓練的物體探測器，並重複
使用其中的一些參數來初始化新模型。

```
# 下載模型
$ wget http://storage.googleapis.com/download.tensorflow.org/models/object_
detection/faster_rcnn_resnet101_coco_11_06_2017.tar.gz
$ tar -xvf faster_rcnn_resnet101_coco_11_06_2017.tar.gz
$ gsutil cp faster_rcnn_resnet101_coco_11_06_2017/model.ckpt.* gs://${YOUR_
GCS_BUCKET}/data/

# 把儲存的路徑轉換到設定檔案中
$ sed -i "s|PATH_TO_BE_CONFIGURED|"gs://${YOUR_GCS_BUCKET}"/data|g" \
    object_detection/samples/configs/faster_rcnn_resnet101_pets.config

# 將設定檔案複製到 Google 儲存中
$ gsutil cp object_detection/samples/configs/faster_rcnn_resnet101_ pets.
config \
    gs://${YOUR_GCS_BUCKET}/data/faster_rcnn_resnet101_pets.config
```

至此，儲存區域裡應該有以下中間結果檔案，如圖 3-12 所示。

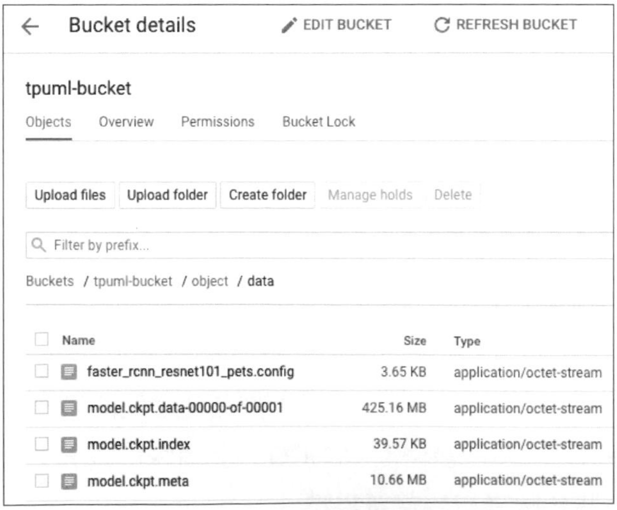

圖 3-12　中間結果檔案

接著，把剛才準備的檔案包裝，執行下面的指令：

```
$ bash object_detection/dataset_tools/create_pycocotools_package.sh /tmp /
pycocotools
$ python setup.py sdist
$ (cd slim && python setup.py sdist)
```

最後，可以執行訓練了。在這裡，訓練和評估（Evaluation）可以同時進行，程式如下：

```
$ gcloud ml-engine jobs submit training `whoami`_object_detection_
pets_`date +%m_%d_%Y_%H_%M_%S` \
    --runtime-version 1.9 \
    --job-dir=gs://${YOUR_GCS_BUCKET}/model_dir \
    --packages dist/object_detection-0.1.tar.gz,slim/dist/slim-0.1.tar.gz, /
tmp/pycocotools/pycocotools-2.0.tar.gz \
    --module-name object_detection.model_main \
    --region us-central1 \
    --config object_detection/samples/cloud/cloud.yml \
    -- \
    --model_dir=gs://${YOUR_GCS_BUCKET}/model_dir \
    --pipeline_config_path=gs://${YOUR_GCS_BUCKET}/data/faster_rcnn_
resnet101_pets.config
```

如圖 3-13 所示顯示了訓練中間結果。

```
worker-replica-0
worker-replica-0 +-------------------------------+----------------------+----------------------+
worker-replica-0 | N/A 41C P0 138W / 149W | 10954MiB / 11441MiB | 81% Default |
worker-replica-0 | 0 Tesla K80 Off | 00000000:00:04.0 Off | 0 |
worker-replica-0 |===============================+======================+======================|
worker-replica-0 | Fan Temp Perf Pwr:Usage/Cap| Memory-Usage | GPU-Util Compute M. |
worker-replica-0 | GPU Name Persistence-M| Bus-Id Disp.A | Volatile Uncorr. ECC |
worker-replica-0 |-------------------------------+----------------------+----------------------+
worker-replica-0 | NVIDIA-SMI 410.79 Driver Version: 410.79 CUDA Version: 10.0 |
worker-replica-0 +-------------------------------+----------------------+----------------------+
```

圖 3-13 訓練中間結果

在訓練的同時，我們可以開啟另外一個 Shell，用 TensorBoard 去檢查訓練的進展即中間結果，如圖 3-14 所示。

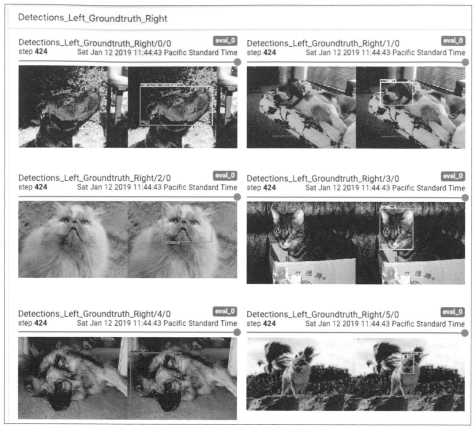

圖 3-14　檢查中間結果

現在就可以使用 TPU 進行訓練了，程式如下：

```
# 重新輸出到一個新的資料夾中
$ MODEL_DIR=tpuml-bucket/object/model2

# 訓練
$ gcloud ml-engine jobs submit training `whoami`_object_detection_`date
+%m_%d_%Y_%H_%M_%S` \
```

```
  --job-dir=gs://${MODEL_DIR} \
  --packages dist/object_detection-0.1.tar.gz,slim/dist/slim-0.1.tar.gz, /
tmp/pycocotools/pycocotools-2.0.tar.gz \
  --module-name object_detection.model_tpu_main \
  --runtime-version 1.9 \
  --scale-tier BASIC_TPU \
  --region us-central1 \
  -- \
  --tpu_zone us-central1 \
  --model_dir=gs://${MODEL_DIR} \
  --pipeline_config_path=gs://${YOUR_GCS_BUCKET}/data/faster_rcnn_
resnet101_pets_tpu.config

# 評價
$ gcloud ml-engine jobs submit training object_detection_eval `date
+%m_%d_%Y_%H_%M_%S` \
  --runtime-version 1.9 \
  --job-dir=gs://${MODEL_DIR} \
  --packages dist/object_detection-0.1.tar.gz,slim/dist/slim-0.1.tar.gz, /
tmp/pycocotools/pycocotools-2.0.tar.gz \
  --module-name object_detection.model_main \
  --region us-central1 \
  --scale-tier BASIC_GPU \
  -- \
  --model_dir=gs://${MODEL_DIR} \
  --pipeline_config_path=gs://${YOUR_GCS_BUCKET}/data/faster_rcnn_
resnet101_pets_tpu.config \
  --checkpoint_dir=gs://${MODEL_DIR}
```

TPU 在 Google 內部已經被廣泛使用，但作為開發者，筆者在使用
Google 雲 TPU 的時候，還是遇到了一些問題，主要是版本相容和一些
文件沒有及時更新的問題。在實際應用的時候，可能要做一些變通，開
發者一般透過公開的通道，可以從 Google 取得答案。

在模型產生之後，我們需要將其轉換成 TensorFlow Lite 能使用的模式。
首先，產生 TensorFlow Lite 的模型檔案，讓我們先執行下面的程式：

```
$ export CONFIG_FILE=gs://tpuml-bucket/object/data/pipeline.config
$ export CHECKPOINT_PATH=gs://tpuml-bucket/object/data/model.ckpt
$ export OUTPUT_DIR=/tmp/tflite

$ object_detection/export_tflite_ssd_graph.py \
    --pipeline_config_path=$CONFIG_FILE \
    --trained_checkpoint_prefix=$CHECKPOINT_PATH \
    --output_directory=$OUTPUT_DIR \
    --add_postprocessing_op=true
```

執行後，產生以下內容：

```
tflite_graph.pb
tflite_graph.pbtxt
```

然後，執行以下程式：

```
$ bazel run --config=opt tensorflow/lite/toco:toco -- \
    --input_file=$OUTPUT_DIR/tflite_graph.pb \
    --output_file=$OUTPUT_DIR/detect.tflite \
    --input_shapes=1,300,300,3 \
    --input_arrays=normalized_input_image_tensor \
    --output_arrays='TFLite_Detection_PostProcess','TFLite_Detection_
PostProcess:1','TFLite_Detection_PostProcess:2','TFLite_Detection_
PostProcess:3' \
    --inference_type=QUANTIZED_UINT8 \
    --mean_values=128 \
    --std_values=128 \
    --change_concat_input_ranges=false \
    --allow_custom_ops
```

上面的程式執行完畢，就會獲得 detect.tflite 檔案，我們可以把這個模型檔案 pet_label_ map.pbtxt 整合進 TensorFlow Lite 的應用，再執行在終端裝置上，實作方式後面會詳細介紹。

關於 TensorFlow Lite 的模型格式和轉換方法，後面也會詳細說明，在這裡只是大概介紹一下。

3.4 使用 TensorFlow Serving 最佳化 TensorFlow 模型

下面筆者來介紹一下如何使用 TensorFlow Serving 元件匯出已經過訓練的 TensorFlow 模型，並使用標準 tensorflow_model_server 來對它提供服務。如果想了解更多資訊，請參閱 TensorFlow Serving 高階教學（https://tensorflow.google.cn/serving/serving_advanced? hl=zh-CN）。

本書使用 TensorFlow 教學中引用的簡單 Softmax 回歸模型進行手寫影像（MNIST 資料）分類。如果開發者不熟悉 TensorFlow 或 MNIST 是什麼，可以參閱 MNIST For ML Beginners 教學（http://tensorflow.google.cn/tutorials/mnist/beginners/index.html?hl=zh-CN #mnist-for-ml-beginners）。

本書的程式由兩部分組成：

一個是 Python 檔案 mnist_saved_model.py（https://github.com/tensorflow/serving/tree/ master/ tensorflow_serving/example/mnist_saved_model.py），用於訓練和匯出模型。

另一個是 ModelServer 二進位檔案，可以使用 Apt 安裝，也可以從 C++ 檔案（main.cc）編譯（https://github.com/tensorflow/serving/tree/master/

tensorflow_serving/model_servers/ main.cc）。TensorFlow Serving 的 ModelServer 會發現新的匯出模型，並執行 gRPC 服務來為其服務（http: //www.grpc.io/）。gRPC 是 Google 開放原始碼的高效輕量級處理程序通訊協定，Google 的對外介面基本都支援這個協定，因此被很多國內網際網路公司所採用。

TensorFlow 模型訓練開始之前，需要安裝 Docker（https://tensorflow. google.cn/serving/ docker?hl=zh-CN#installing-docker）。

3.4.1 訓練和匯出 TensorFlow 模型

正如在 mnist_saved_model.py 中所見，訓練的方式與在初學者 MNIST 教學中完成的方式相同（https://tensorflow.google.cn/get_started/mnist/ beginners?hl=zh-CN）。TensorFlow 執行圖在 TensorFlow 階段中啟動，輸入張量（影像）為 x，輸出張量（Softmax 得分）為 y。

然後，我們使用 TensorFlow 的 SavedModelBuilder 模組匯出模型，程式如下：

```
"""SavedModel builder.
Builds a SavedModel that can be saved to storage, is language neutral, and
enables systems to produce, consume, or transform TensorFlow Models.
"""


from __future__ import absolute_import
from __future__ import division
from __future__ import print_function

from tensorflow.python.saved_model.builder_impl import _SavedModelBuilder
from tensorflow.python.saved_model.builder_impl import SavedModelBuilder
```

SavedModelBuilder 將訓練模型的「快照」儲存到可靠儲存中，以便稍後載入並進行推理。有關 SavedModel 格式的詳細資訊，請參閱 SavedModel 中 的 README.md 文 件（https://github.com/tensorflow/tcnsorflow/blob/master/tensorflow/python/saved_model/READ- M E.md）。

下面的程式（https://github.com/tensorflow/serving/tree/master/tensorflow_serving/exam- ple/mnist_saved_model.py）可以訓練和匯出 minst 模型：

```python
#! /usr/bin/env python
r"""Train and export a simple Softmax Regression TensorFlow model.
The model is from the TensorFlow "MNIST For ML Beginner" tutorial. This program
simply follows all its training instructions, and uses TensorFlow SavedModel to
export the trained model with proper signatures that can be loaded by standard
tensorflow_model_server.
Usage: mnist_saved_model.py [--training_iteration=x] [--model_version=y] \
    export_dir
"""

from __future__ import print_function

import os
import sys

import tensorflow as tf

import mnist_input_data

tf.app.flags.DEFINE_integer('training_iteration', 1000,
                            'number of training iterations.')
tf.app.flags.DEFINE_integer('model_version', 1, 'version number of the model.')
tf.app.flags.DEFINE_string('work_dir', '/tmp', 'Working directory.')
FLAGS = tf.app.flags.FLAGS
```

```
def main(_):
  if len(sys.argv) < 2 or sys.argv[-1].startswith('-'):
    print('Usage: mnist_saved_model.py [--training_iteration=x] '
          '[--model_version=y] export_dir')
    sys.exit(-1)
  if FLAGS.training_iteration <= 0:
    print('Please specify a positive value for training iteration.')
    sys.exit(-1)
  if FLAGS.model_version <= 0:
    print('Please specify a positive value for version number.')
    sys.exit(-1)

  # 訓練模型
  print('Training model...')
  mnist = mnist_input_data.read_data_sets(FLAGS.work_dir, one_hot=True)
  sess = tf.InteractiveSession()
  serialized_tf_example = tf.placeholder(tf.string, name='tf_example')
  feature_configs = {'x': tf.FixedLenFeature(shape=[784], dtype=
tf.float32),}
  tf_example = tf.parse_example(serialized_tf_example, feature_configs)
  x = tf.identity(tf_example['x'], name='x')  # use tf.identity() to assign
name
  y_ = tf.placeholder('float', shape=[None, 10])
  w = tf.Variable(tf.zeros([784, 10]))
  b = tf.Variable(tf.zeros([10]))
  sess.run(tf.global_variables_initializer())
  y = tf.nn.softmax(tf.matmul(x, w) + b, name='y')
  cross_entropy = -tf.reduce_sum(y_ * tf.log(y))
  train_step = tf.train.GradientDescentOptimizer(0.01).minimize(cross_ entropy)
  values, indices = tf.nn.top_k(y, 10)
  table = tf.contrib.lookup.index_to_string_table_from_tensor(
      tf.constant([str(i) for i in range(10)]))
```

```
prediction_classes = table.lookup(tf.to_int64(indices))
for _ in range(FLAGS.training_iteration):
  batch = mnist.train.next_batch(50)
  train_step.run(feed_dict={x: batch[0], y_: batch[1]})
correct_prediction = tf.equal(tf.argmax(y, 1), tf.argmax(y_, 1))
accuracy = tf.reduce_mean(tf.cast(correct_prediction, 'float'))
print('training accuracy %g' % sess.run(
    accuracy, feed_dict={
        x: mnist.test.images,
        y_: mnist.test.labels
    }))
print('Done training!')

# 匯出模型
  export_path_base = sys.argv[-1]
export_path = os.path.join(
    tf.compat.as_bytes(export_path_base),
    tf.compat.as_bytes(str(FLAGS.model_version)))
print('Exporting trained model to', export_path)
builder = tf.saved_model.builder.SavedModelBuilder(export_path)

# 建立 signature_def_map
classification_inputs = tf.saved_model.utils.build_tensor_info(
    serialized_tf_example)
classification_outputs_classes = tf.saved_model.utils.build_
tensor_info(
    prediction_classes)
classification_outputs_scores = tf.saved_model.utils.build_tensor_
info(values)

classification_signature = (
    tf.saved_model.signature_def_utils.build_signature_def(
        inputs={
```

```
                tf.saved_model.signature_constants.CLASSIFY_INPUTS:
                    classification_inputs
            },
            outputs={
                tf.saved_model.signature_constants.CLASSIFY_OUTPUT_CLASSES:
                    classification_outputs_classes,
                tf.saved_model.signature_constants.CLASSIFY_OUTPUT_SCORES:
                    classification_outputs_scores
            },
            method_name=tf.saved_model.signature_constants.CLASSIFY_METHOD_
NAME))

    tensor_info_x = tf.saved_model.utils.build_tensor_info(x)
    tensor_info_y = tf.saved_model.utils.build_tensor_info(y)

    prediction_signature = (
        tf.saved_model.signature_def_utils.build_signature_def(
            inputs={'images': tensor_info_x},
            outputs={'scores': tensor_info_y},
            method_name=tf.saved_model.signature_constants.PREDICT_METHOD_ NAME))

    builder.add_meta_graph_and_variables(
        sess, [tf.saved_model.tag_constants.SERVING],
        signature_def_map={
            'predict_images':
                prediction_signature,
            tf.saved_model.signature_constants.DEFAULT_SERVING_SIGNATURE_ DEF_
KEY:
                classification_signature,
        },
        main_op=tf.tables_initializer(),
        strip_default_attrs=True)
```

```
  builder.save()

  print('Done exporting!')

if __name__ == '__main__':
  tf.app.run()
```

下面這個簡短的通用程式片段的功能是將模型儲存到機器磁碟。

```
export_path_base = sys.argv[-1]

export_path = os.path.join(
    compat.as_bytes(export_path_base),
    compat.as_bytes(str(FLAGS.model_version)))
print 'Exporting trained model to', export_path
builder = tf.saved_model.builder.SavedModelBuilder(export_path)
builder.add_meta_graph_and_variables(
    sess, [tag_constants.SERVING],
    signature_def_map={
      'predict_images':
          prediction_signature,
      signature_constants.DEFAULT_SERVING_SIGNATURE_DEF_KEY:
          classification_signature,
    },
    main_op=main_op)

  builder.save()
```

SavedModelBuilder 採用以下參數：

export_path 匯出目錄的路徑。如果目錄不存在，SavedModelBuilder 將建立該目錄。在範例中，我們連接命令列參數和 FLAGS.model_version 以取得匯出目錄。

FLAGS.model_version 指定模型的版本。在匯出同一模型的較新版本時，應指定較大的整數值。每個版本將匯出到指定路徑下的不同子目錄中。

開發者可以透過 SavedModelBuilder.add_meta_graph_and_variables() 函數將元圖和變數增加到建置元中。該函數的相關參數解釋如下：

sess TensorFlow 階段，包含正在匯出的訓練模型。

tags 用於儲存元圖的標記集。由於我們打算在服務中使用圖形，因此我們使用來自預先定義的 SavedModel 標記常數的 serve 標記。更多詳細資訊，請參閱 tag_constants.py（https://github.com/tensorflow/tensorflow/blob/master/tensorflow/python/saved_model/tag_constants.py）和相關的 TensorFlow API 文件（https://tensorflow.google.cn/api_docs/python/tf/ saved_model/tag_constants?hl=zh-CN）。

signature_def_map 將指定使用者提供的簽名對映到 tensorflow::SignatureDef，Signature 指定正在匯出的模型類型，以及在執行推理時綁定的輸入 / 輸出張量。

serving_default 特殊簽名金鑰，用於指定預設服務簽名。預設服務簽名 def 鍵及與簽名相關的其他常數被定義為 SavedModel 簽名常數的一部分。更多詳細資訊請參閱 signature_constants.py（https://github.com/tensorflow/tensorflow/blob/master/tensorflow/python/ saved_model/ signature_constants.py）和相關的 TensorFlow 1.0 API 文件（https://tensorflow. google.cn/api_docs/ python/tf/saved_model/signature_constants?hl= zh-CN）。

此外，為了建置簽名定義，SavedModel API 提供了 Module: tf.saved_model.signature_ def_utils（https://tensorflow.google.cn/api_docs/python/tf/saved_model/signature_def_utils? hl=zh-CN）具體地説，在原始的 mnist_ saved_model.py 檔 案 中（https://github.com/tensorflow/ serving/tree/ master/ tensorflow_serving/example/mnist_saved_model.py），我們將 signature_def_utils. build_signature_def( ) 用於 buildpredict_signature 和 classification_signature。

那麼，如何使用 predict_signature 呢？相關參數解釋如下：

inputs={'images': tensor_info_x} 指定輸入張量資訊。
outputs={'scores': tensor_info_y} 指定分數張量資訊。

method_name 是 用 於 推 理 的 方 法。 對 於 預 測 請 求， 應 將 其 設 定 為 tensorflow/serving/predict。有關其他方法名稱，請參閱 signature_constants. py（https://github.com/tensorflow/tensorflow/blob/master/tensorflow/python/ saved_model/signature_constants.py） 和 relatedTensorFlow 1.0 API 文 件 （https://tensorflow.google.cn/api_docs/python/tf/saved_model/ signature_ constants? hl=zh-CN）。

請 注 意，tensor_info_x 和 tensor_info_y 具 有 此 處 定 義 的 tensorflow:: TensorInfo 協 定 緩 衝 區 的 結 構（https://github.com/tensorflow/tensorflow/ blob/master/tensorflow/core/protobuf/ meta_graph. proto）。

為了建置張量資訊，TensorFlow SavedModel API 還提供了 utils.py，程 式如下：

```
# Copyright 2016 The TensorFlow Authors. All Rights Reserved.
#
# Licensed under the Apache License, Version 2.0 (the "License");
# you may not use this file except in compliance with the License.
# You may obtain a copy of the License at
#
#     http://www.apache.org/licenses/LICENSE-2.0
#
# Unless required by applicable law or agreed to in writing, software
```

```
# distributed under the License is distributed on an "AS IS" BASIS,
# WITHOUT WARRANTIES OR CONDITIONS OF ANY KIND, either express or implied.
# See the License for the specific language governing permissions and
# limitations under the License.
# ==========================================================================
"""SavedModel utility functions.
Utility functions to assist with setup and construction of the SavedModel
proto.
"""
from __future__ import absolute_import
from __future__ import division
from __future__ import print_function

# pylint: disable=unused-import
from tensorflow.python.saved_model.utils_impl import build_tensor_info
from tensorflow.python.saved_model.utils_impl import build_tensor_info_
from_op
from tensorflow.python.saved_model.utils_impl import get_tensor_from_
tensor_info
# pylint: enable=unused-import
```

原始程式碼的位址為 https://github.com/tensorflow/tensorflow/blob/master/
tensorflow/python/ saved_model/utils.py，相 關 文 件 的 位 址 為 https://
tensorflow.google.cn/api_docs/python/ tf/saved_model/utils?hl=zh-CN。

另外一點需要注意的是，影像和分數是張量別名，它們可以是開發者想
要的任何特定的字串。作為張量 x 和 y 的邏輯名稱，可以在稍後發送預
測請求時參考張量進行綁定。

舉例來說，如果 x 參考名稱為 "long_tensor_name_foo" 的張量，y 參考名
稱為 "generated_ tensor_name_bar" 的張量，則建置元將張量邏輯名稱儲
存為實名對映：

```
'images' -> 'long_tensor_name_foo'
'score' -> 'generated_tensor_name_bar'
```

使用者在執行推理時可以使用其邏輯名稱來參考這些張量。

下面就可以執行了。

首先,複製原始程式到本機:

```
$ git clone https://github.com/tensorflow/serving.git
$ cd serving
```

然後,清除輸出目錄,結果會輸出到這個目錄中:

```
rm -rf /tmp/mnist
```

現在,讓我們訓練模型:

```
$ tools/run_in_docker.sh python tensorflow_serving/example/mnist_saved_
model.py /tmp/mnist
```

執行過程的螢幕輸出如下:

```
Training model...

...


Done training!
Exporting trained model to models/mnist
Done exporting!
```

檢視匯出的目錄:

```
$ ls /tmp/mnist

1
```

如上所述，為了匯出模型的每個版本，將建立一個子目錄。FLAGS.
model_version 的預設值為 1，因此建立了對應的子目錄 1，程式如下：

```
$ ls /tmp/mnist/1
saved_model.pb variables
```

每個版本的子目錄下都包含 saved_model.pb 和 variables 兩個檔案。
saved_model.pb 是序列化的 tensorflow::SavedModel，它包含模型的或多
個圖形定義，以及模型的中繼資料（如簽名）；Variable 是包含圖形序列
化變數的檔案。

3.4.2 使用標準 TensorFlow ModelServer 載入匯出 的模型

使用 Docker 服務載入模型，程式如下：

```
$ docker run -p 8500:8500 \
  --mount type=bind,source=/tmp/mnist,target=/models/mnist \
  -e MODEL_NAME=mnist -t tensorflow/serving &
```

3.4.3 測試伺服器

我們可以使用下面的 mnist_client 程式（原始程式碼在 https://github.com/
tensorflow/ serving/tree/master/tensorflow_serving/example/mnist_client.
py）來測試伺服器，該用戶端程式下載 MNIST 測試資料，將其作為請
求發送給伺服器，並計算推理錯誤率。

```
#!/usr/bin/env python2.7


"""A client that talks to tensorflow_model_server loaded with mnist model.
The client downloads test images of mnist data set, queries the service with
```

```
such test images to get predictions, and calculates the inference error rate.
Typical usage example:
    mnist_client.py --num_tests=100 --server=localhost:9000
"""


from __future__ import print_function

import sys
import threading

import grpc
import numpy
import tensorflow as tf

from tensorflow_serving.apis import predict_pb2
from tensorflow_serving.apis import prediction_service_pb2_grpc
import mnist_input_data

tf.app.flags.DEFINE_integer('concurrency', 1,
                            'maximum number of concurrent inference requests')
tf.app.flags.DEFINE_integer('num_tests', 100, 'Number of test images')
tf.app.flags.DEFINE_string('server', '', 'PredictionService host:port')
tf.app.flags.DEFINE_string('work_dir', '/tmp', 'Working directory. ')
FLAGS = tf.app.flags.FLAGS

class _ResultCounter(object):
  """Counter for the prediction results."""

  def __init__(self, num_tests, concurrency):
    self._num_tests = num_tests
    self._concurrency = concurrency
    self._error = 0
    self._done = 0
```

```
      self._active = 0
      self._condition = threading.Condition()

  def inc_error(self):
    with self._condition:
      self._error += 1

  def inc_done(self):
    with self._condition:
      self._done += 1
      self._condition.notify()

  def dec_active(self):
    with self._condition:
      self._active -= 1
      self._condition.notify()

  def get_error_rate(self):
    with self._condition:
      while self._done != self._num_tests:
        self._condition.wait()
      return self._error / float(self._num_tests)

  def throttle(self):
    with self._condition:
      while self._active == self._concurrency:
        self._condition.wait()
      self._active += 1

def _create_rpc_callback(label, result_counter):
  """Creates RPC callback function.
  Args:
    label: The correct label for the predicted example.
```

```
        result_counter: Counter for the prediction result.
    Returns:
        The callback function.
    """
    def _callback(result_future):
        """Callback function.
        Calculates the statistics for the prediction result.
        Args:
            result_future: Result future of the RPC.
        """
        exception = result_future.exception()
        if exception:
            result_counter.inc_error()
            print(exception)
        else:
            sys.stdout.write('.')
            sys.stdout.flush()
            response = numpy.array(
                result_future.result().outputs['scores'].float_val)
            prediction = numpy.argmax(response)
            if label != prediction:
                result_counter.inc_error()
        result_counter.inc_done()
        result_counter.dec_active()
    return _callback


def do_inference(hostport, work_dir, concurrency, num_tests):
    """Tests PredictionService with concurrent requests.
    Args:
        hostport: Host:port address of the PredictionService.
        work_dir: The full path of working directory for test data set.
        concurrency: Maximum number of concurrent requests.
        num_tests: Number of test images to use.
```

```
  Returns:
    The classification error rate.
  Raises:
    IOError: An error occurred processing test data set.
  """
  test_data_set = mnist_input_data.read_data_sets(work_dir).test
  channel = grpc.insecure_channel(hostport)
  stub = prediction_service_pb2_grpc.PredictionServiceStub(channel)
  result_counter = _ResultCounter(num_tests, concurrency)
  for _ in range(num_tests):
    request = predict_pb2.PredictRequest()
    request.model_spec.name = 'mnist'
    request.model_spec.signature_name = 'predict_images'
    image, label = test_data_set.next_batch(1)
    request.inputs['images'].CopyFrom(
        tf.contrib.util.make_tensor_proto(image[0], shape=[1, image[0].size]))
    result_counter.throttle()
    result_future = stub.Predict.future(request, 5.0)  # 5 seconds
    result_future.add_done_callback(
        _create_rpc_callback(label[0], result_counter))
  return result_counter.get_error_rate()


def main(_):
  if FLAGS.num_tests > 10000:
    print('num_tests should not be greater than 10k')
    return
  if not FLAGS.server:
    print('please specify server host:port')
    return
  error_rate = do_inference(FLAGS.server, FLAGS.work_dir,
                            FLAGS.concurrency, FLAGS.num_tests)
  print('\nInference error rate: %s%%' % (error_rate * 100))
```

```
if __name__ == '__main__':
  tf.app.run()
$ tools/run_in_docker.sh python tensorflow_serving/example/mnist_ client.py \
  --num_tests=1000 --server=127.0.0.1:8500
```

程式輸出結果如下：

```
Inference error rate: 11.13%
```

對於訓練好的 Softmax 模型，我們預計準確率約為 90%，前 1000 個測試影像的推理錯誤率為 11.13%。從這個結果，我們就可以確認伺服器成功載入並執行了已訓練的模型！

3.5 TensorFlow 擴充（Extended）

在機器學習裡，我們非常關注模型程式，但是 TensorFlow Extended 不只是模型。那麼，TensorFlow Extended 解決了哪些問題呢？

機器學習的程式其實很簡單，但資料收集、設定、機器管理等事情需要投入大量的時間和精力，那麼這些需要花費週邊力量的工作，我們是否能夠透過其他方式來實現，進一步讓專案快速實施落實呢？

TensorFlow 擴充（Extended）就是 Google 推出的能夠幫助開發者解決這些問題、實現專案快速實施落實的有效工具。TensorFlow 擴充是 Google 內部廣泛使用的基礎架構，是機器學習平台的主要組成部分。但是，現在還有很多程式和功能沒有開放原始碼。這個架構依靠於 Google 強大的基礎設施，也是 Google 強大機器學習能力的一部分。

建置
TensorFlow Mobile

本章主要介紹如何建置 TensorFlow Mobile 的結構，以及如何建置應用。

4.1 TensorFlow Mobile 的歷史

TensorFlow Mobile 是 TensorFlow 的第一個對行動端嵌入式裝置的支援架構。在早期的 TensorFlow 發佈中就加入了對包含 Android、iOS 和樹莓派的支援，它主要採用交換編譯的方式，能夠縮短開發週期，並在裝置上快速執行。

4.2 TensorFlow 程式結構

讓我們先下載 TensorFlow 的原始程式碼。我們可以使用 Git 複製專案：

```
$ git clone https://github.com/tensorflow/tensorflow.git
```

針對 TensorFlow 開發人員，可以使用 Git 檢視歷史記錄並管理分支。還可以下載包含 TensorFlow 原始程式碼的壓縮檔（https://github.com/tensorflow/tensorflow/archive/master.zip）。

讓我們來看一下 TensorFlow 的程式結構：

```
-rw-rw-r-- BUILD
drwxrwxr-x c
drwxrwxr-x cc
-rw-rw-r-- .clang-format
-rw-rw-r-- .blazerc
drwxrwxr-x compiler
drwxrwxr-x contrib
drwxrwxr-x core
drwxrwxr-x docs_src
drwxrwxr-x examples
drwxrwxr-x g3doc
drwxrwxr-x go
-rw-rw-r-- __init__.py
-rw-rw-r-- __init__.pyc
drwxrwxr-x java
drwxrwxr-x python
drwxrwxr-x stream_executor
-rw-rw-r-- tensorflow.bzl
-rw-rw-r-- tf_exported_symbols.lds
-rw-rw-r-- tf_version_script.lds
drwxrwxr-x tools
```

```
drwxrwxr-x user_ops
-rw-rw-r-- version_check.bzl
-rw-rw-r-- workspace.bzl
```

從上面程式可以看到，根目錄中有兩個重要的資料夾，一個是包含原始程式碼的資料夾 core，另一個是 tools 資料夾。還有兩個檔案——.blazerc 和在 tools 資料夾下的 tf_env_collect.sh。其中 .bazelrc 定義了 Bazel 設定，tf_env_collect.sh 用於收集系統資訊，可以在提交 bug 時附加結果。

以下資料夾包含了 TensorFlow 實現，它支援 C、C++、Go、Java、Python。

```
drwxrwxr-x c
drwxrwxr-x cc
drwxrwxr-x compiler
drwxrwxr-x core
drwxrwxr-x go
drwxrwxr-x java
drwxrwxr-x python
```

docs_src 包含了文件檔案：

```
drwxrwxr-x docs_src
```

examples 包含應用的實例，在這些實例中也包含 Android、iOS 的應用。

```
drwxrwxr-x examples
```

contrib 是一個特別的資料夾，它的作用在 README.md 裡有説明：

Any code in this directory is not officially supported, and may change or be removed at any time without notice.

這段文字表明,此資料夾中的任何程式都不是官方的。如果想在產品中使用,就必須承擔維護程式的責任,使其能夠相容。

在 contrib 中,有兩個我們將頻繁使用的資料夾。一個是 android,它是 TensorFlow Android 範例應用程式;另一個是 lite,它是 TensorFlow Lite。有趣的是,Java 不在 contrib 資料夾中。

Java 不在 contrib 資料夾中,是因為官方支援 TensorFlow 的 Java 介面。在 contrib/android 中的 TensorFlow Lite 和黏合程式不受官方支援。我們需要牢記這些。它現在可能不是很重要,但它可能會影響開發者未來的開發計畫。

讓我們來看一段 TensorFlow API 文件中的話:

TensorFlow has APIs available in several languages both for constructing and executing a TensorFlow graph. The Python API is at present the most complete and the easiest to use, but other language APIs may be easier to integrate into projects and may offer some performance advantages in graph execution.

A word of caution: the APIs in languages other than Python are not yet covered by the API stability promises.

由此可見,官方只支援 Python API。

我們先看看底層實現。TensorFlow 的底層都是利用 C 和 C++ 實現的。// tensorflow/ cc:cc_ops 定義了自動產生的 C++ 介面,程式如下:

```
tf_gen_op_wrappers_cc(
    name = "cc_ops",
    api_def_srcs = ["//tensorflow/core/api_def:base_api_def"],
```

```
    op_lib_names = [
        "array_ops",
        "audio_ops",
        "candidate_sampling_ops",
        "control_flow_ops",
        "data_flow_ops",
        "image_ops",
        "io_ops",
        "linalg_ops",
        "logging_ops",
        "lookup_ops",
        "manip_ops",
        "math_ops",
        "nn_ops",
        "no_op",
        "parsing_ops",
        "random_ops",
        "sparse_ops",
        "state_ops",
        "string_ops",
        "training_ops",
        "user_ops",
    ],
    other_hdrs = [
        "ops/const_op.h",
        "ops/standard_ops.h",
    ],
    pkg = "//tensorflow/core",
)
```

TensorFlow C++ 參考中也清楚地記錄了這些 Ops（Operations, 運算元）。
TensorFlow 的 API 文件也是從程式中分析的。

讓我們看一下 TensorFlow Android 的示範程式：

```
tf_cuda_library(
    name = "native",
    srcs = glob(["*.cc"]) + select({
        # The Android toolchain makes "jni.h" available in the include path.
        # For non-Android toolchains, generate jni.h and jni_md.h.
        "//tensorflow:android": [],
        "//conditions:default": [
            ":jni.h",
            ":jni_md.h",
        ],
    }),
    hdrs = glob(["*.h"]),
    copts = tf_copts() + [
        "-landroid",
        "-llog",
    ],
    includes = select({
        "//tensorflow:android": [],
        "//conditions:default": ["."],
    }),
    linkopts = [
        "-landroid",
    "-llog",
    ],
    deps = [
        **"//tensorflow/c:c_api",**
    ] + select({
        "//tensorflow:android": [
            **"//tensorflow/core:android_tensorflow_lib",**
        ],
        "//conditions:default": [
            "//tensorflow/core:all_kernels",
```

```
            "//tensorflow/core:direct_session",
            "//tensorflow/core:ops",
        ],
    }),
    alwayslink = 1,
)
//tensorflow/c:c_api 的定義
tf_cuda_library(
    name = "c_api",
    srcs = [
        "c_api.cc",
        "c_api_function.cc",
    ],
    hdrs = [
        "c_api.h",
    ],
    copts = tf_copts(),
    visibility = ["//visibility:public"],
    deps = select({
        "//tensorflow:android": [
            ":c_api_internal",
            "//tensorflow/core:android_tensorflow_lib_lite",
        ],
        "//conditions:default": [
        ],
    }) + select({
        "//tensorflow:with_xla_support": [
            "//tensorflow/compiler/tf2xla:xla_compiler",
            "//tensorflow/compiler/jit",
        ],
        "//conditions:default": [],
    }),
)
//tensorflow/core:android_tensorflow_lib 的定義
```

```
cc_library(
    name = "android_tensorflow_lib",
    srcs = if_android([":android_op_registrations_and_gradients"]),
    copts = tf_copts(),
    tags = [
        "manual",
        "notap",
    ],
    visibility = ["//visibility:public"],
    deps = [
        ":android_tensorflow_lib_lite",
        ":protos_all_cc_impl",
        "//tensorflow/core/kernels:android_tensorflow_kernels",
        "//third_party/eigen3",
        "@protobuf_archive//:protobuf",
    ],
    alwayslink = 1,
)
```

//tensorflow/core/kernels:android_tensorflow_kernels 基本包含了 Ops 的實現,它包含 "//tensorflow/core/kernels:android_core_ops" 和 "//tensorflow/core/kernels:android_extended_ops"

```
cc_library(
    name = "android_tensorflow_kernels",
    srcs = select({
        "//tensorflow:android": [
            "//tensorflow/core/kernels:android_core_ops",
            "//tensorflow/core/kernels:android_extended_ops",
        ],
        "//conditions:default": [],
    }),
    copts = tf_copts(),
    linkopts = select({
        "//tensorflow:android": [
            "-ldl",
```

```
    ],
    "//conditions:default": [],
}),
tags = [
    "manual",
    "notap",
],
visibility = ["//visibility:public"],
deps = [
    "//tensorflow/core:android_tensorflow_lib_lite",
    "//tensorflow/core:protos_all_cc_impl",
    "//third_party/eigen3",
    "//third_party/fft2d:fft2d_headers",
    "@fft2d",
    "@gemmlowp",
    "@protobuf_archive//:protobuf",
],
alwayslink = 1,
)
```

這個示範程式的建置目標是 //tensorflow/examples/android:tensorflow_
demo，這個建置目標不僅依賴於 Java JNI，而且也依賴於 Java 的原生實
現（//tensorflow/java/src/main/native: native）。

4.3 建置及執行

在 1.10 版本之前，開發者需要手動編輯 WORKSPACE 設定工具鏈，程
式如下：

```
android_sdk_repository(
    name = "androidsdk",
```

```
    api_level = 23,
    # 確保你在 SDK 管理員中安裝了 build_tools_version，因為它會定期更新
    build_tools_version = "26.0.1",
    # 取代為系統中 Android SDK 的路徑
    **path = "/opt/Android/sdk",**
)

android_ndk_repository(
    name="androidndk",
    **path="/opt/Android/ndk",**
    # 編譯 TensorFlow 需要將 api_level 設定為 14 或更高
    # 請注意，NDK 版本不是 API 等級
    api_level=14)
```

在新的版本裡，這個手動的過程被改變了，解決辦法是執行 ./configure，
這個指令稿會提示開發者設定環境變數：

```
Please specify the location of python. [Default is /usr/bin/python]:

Do you wish to build TensorFlow with XLA JIT support? [Y/n]:
Do you wish to build TensorFlow with OpenCL SYCL support? [y/N]:
Do you wish to build TensorFlow with ROCm support? [y/N]:
Do you wish to build TensorFlow with CUDA support? [y/N]:
Do you wish to download a fresh release of clang? (Experimental) [y/N]:
Do you wish to build TensorFlow with MPI support? [y/N]:
Please specify optimization flags to use during compilation when bazel
option "--config=opt" is specified [Default is -march=native -Wno-sign-
compare]:
Would you like to interactively configure ./WORKSPACE for Android builds?
[y/N]:
```

開發者只需要簡單地回答問題。對於 Android，關鍵是設定工具鏈和 API
Level，我們可以直接設定環境變數，configure 指令稿用於讀取這些變
數，程式如下：

```
export ANDROID_NDK_HOME=/opt/Android/ndk
export ANDROID_SDK_HOME=/opt/Android/sdk
export ANDROID_NDK_API_LEVEL=26
export ANDROID_SDK_API_LEVEL=26
export ANDROID_BUILD_TOOLS_VERSION=26.0.1
```

API Level 設成 26 即可支援 Android NNAPI。Android 的 SDK 中包含 NDK，可以直接用或下載單獨的 NDK。另外，如果是公司的開發者，最好使用統一版本的 SDK 和 NDK。筆者見過一些公司，對此沒有嚴格的限制，這其實存在潛在的危險。舉例來說，NDK r14 產生的程式會有不充分最佳化的地方。筆者在完成從 r12 到 r16 的遷移時，將 SDK 和 NDK 安裝在一個固定的路徑。這樣做有兩個好處，一是可以實現遠端自動管理，二是可以實現跨平台統一管理。

註：configure 不會安裝 SDK 和 NDK，我們可以使用 Android Studio 中的 SDK Manager 來安裝 SDK 和 NDK。

完成這些之後，我們就可以建置應用了，程式如下：

```
$ bazel build -c opt --config=android --cpu=arm64-v8a //tensorflow/
examples/android:tensorflow_demo
$ adb install -r bazel-bin/tensorflow/examples/android/tensorflow_demo.apk
```

上面程式中的參數解釋如下：

-c opt 讓編譯器最佳化程式。

--config=android 設定交換編譯器。最初，Bazel 僅在主機系統和目標系統相同的開發環境中進行編譯。但是，對於嵌入式系統和行動系統，我們必須交換編譯程式。具體地說，開發者的工作站是 Intel/x86，但是我們想要產生的是可以執行 ARM 處理器的二進位碼。

--cpu=arm64-v8a 指定行動裝置的 ABI。

你也可以使用更短的設定，下面來看看這個設定檔案 tools/bazel.rc：

```
# Bazel 需要將 --cpu 和 --fat_apk_cpu 都設定為
# target CPU, 以正確建置瞬態依賴項。讀者可以參考
# https://docs.bazel.build/versions/master/user-manual.html
#flag--fat_apk_cpu
build:android --crosstool_top=//external:android/crosstool
build:android --host_crosstool_top=@bazel_tools//tools/cpp:toolchain
build:android --config=cross_compile
build:android_arm --config=android
build:android_arm --cpu=armeabi-v7a
build:android_arm --fat_apk_cpu=armeabi-v7a
build:android_arm64 --config=android
build:android_arm64 --cpu=arm64-v8a
build:android_arm64 --fat_apk_cpu=arm64-v8a
```

在筆者看來，開發者還可以使用 android_arm64 來實現相同的目標。如何找到用於開發手機的 abi 呢？

有兩種方法，如果你在手機上啟用了 USB 偵錯，則可以執行指令：

```
adb shell getprop | grep abi
```

你應該可以看到以下結果：

```
[ro.product.cpu.abilist]: [arm64-v8a,armeabi-v7a,armeabi]
[ro.product.cpu.abilist32]: [armeabi-v7a,armeabi]
[ro.product.cpu.abilist64]: [arm64-v8a]
```

最近的 Android 版本已停止支援 armeabi，對於不了解技術細節的使用者，Android 建置並使用了所謂「胖」（Fat）APK 來解決這個問題。在「胖」APK 中，該軟體套件包含所有平台的所有可執行程式，這表示在「胖」APK 中，它將包含 arm64-v8a、armeabi-v7a 甚至 Intel/x86 和 MIPS 的程式。

但是，它會使應用程式套件更大。它不僅會導致儲存問題，還需要更多空間來儲存 APK，更大的套件也表示啟動時間慢，消耗更多的電池電量。事實是，在 ARM 裝置上，使用者永遠不會執行 Intel/x86 程式。因此，當我們從 Google 商店（Google Play Store）或其他 APK 商店下載 APK 時，這些服務將首先檢測使用者手機的 abi，然後將所謂的「瘦」APK 發送到我們的裝置。「瘦」APK 僅包含一個 abi 的程式。所以請記住，如果開發者為一個 abi 撰寫最佳化程式，則必須為其他 abi 準備類似的最佳化。

接下來，安裝 APK：

```
$ adb install bazel-bin/tensorflow/examples/android/tensorflow_demo.apk
```

現在，在連接到桌面的手機上，我們會看到一個圖示，點擊該圖示可以執行它。如果有多台裝置連接到主機，則需要按序號選擇裝置。舉例來說，列出連接到主機的裝置：

```
$ adb devices
List of devices attached
adb server is out of date.  killing...
* daemon started successfully *
  UYT0217C05001756device
```

透過下面的指令安裝示範程式：

```
$ adb install -s UYT0217C05001756device bazel-bin/tensorflow/examples/
android/tensorflow_demo.apk
```

如果你喜歡使用 Android Studio 中的即時執行功能，Bazel 也有類似的功能，稱為行動安裝。我們不必執行 Bazel 建置和安裝，可以執行下面的指令：

```
$ adb mobile-install -s UYT0217C05001756device bazel-bin/tensorflow/
examples/android/tensorflow_demo.apk
```

mobile-install 從以下三個方面來改善建置和安裝:

- Sharded dexing。在建置應用程式的 Java 程式後,Bazel 將類別檔案分為大小大致相等的許多片,並在各個分單晶片單獨呼叫 dex。在上次建置後未更改的分片可以不呼叫 dex。

- 增量檔案傳輸。Android 資源,.dex 檔案和本機函數庫將從主 .apk 中刪除,並儲存在單獨的行動安裝目錄下。這樣可以獨立更新程式和 Android 資源,而不需要重新安裝整個應用程式。因此,傳輸檔案所花費的時間更少,只有已更改的 .dex 檔案才會在裝置上重新編譯。

- 從 APK 外部載入應用程式的部分內容。將一個小的存根(stub)應用程式放入 APK 中,從裝置上的行動安裝目錄載入 Android 資源、Java 程式和本機程式,然後將控制權傳輸給實際的應用程式。除下面描述的一些極端情況外,這對應用程式來說都是透明的。

建議開發者使用這個功能。在安裝完成後,你應該會在手機上看到四個應用圖示,如圖 4-1 所示。

現在讓我們看一下 examples/android/ AndroidManifest.xml 這個檔案:

圖 4-1 應用安裝

```
<activity android:name="org.tensorflow.demo.ClassifierActivity"
        android:screenOrientation="portrait"
        android:label="@string/activity_name_classification">
    <intent-filter>
```

```
        <action android:name="android.intent.action.MAIN" />
        <category android:name="android.intent.category.LAUNCHER" />
        <category android:name="android.intent.category.LEANBACK_LAUNCHER" />
    </intent-filter>
</activity>

<activity android:name="org.tensorflow.demo.DetectorActivity"
        android:screenOrientation="portrait"
        android:label="@string/activity_name_detection">
    <intent-filter>
        <action android:name="android.intent.action.MAIN" />
        <category android:name="android.intent.category.LAUNCHER" />
        <category android:name="android.intent.category.LEANBACK_LAUNCHER" />
    </intent-filter>
</activity>

<activity android:name="org.tensorflow.demo.StylizeActivity"
        android:screenOrientation="portrait"
        android:label="@string/activity_name_stylize">
    <intent-filter>
        <action android:name="android.intent.action.MAIN" />
        <category android:name="android.intent.category.LAUNCHER" />
        <category android:name="android.intent.category.LEANBACK_LAUNCHER" />
    </intent-filter>
</activity>

<activity android:name="org.tensorflow.demo.SpeechActivity"
    android:screenOrientation="portrait"
    android:label="@string/activity_name_speech">
    <intent-filter>
        <action android:name="android.intent.action.MAIN" />
        <category android:name="android.intent.category.LAUNCHER" />
        <category android:name="android.intent.category.LEANBACK_LAUNCHER" />
```

```
    </intent-filter>
</activity>
```

從上面程式中可以了解到，一次安裝指令安裝了四個應用程式，TF Stylize、TF Detect、TF Speech、TF Classify。我們可以手動點擊圖示來啟動應用，也可以執行類似的操作來啟動應用：

```
$ adb shell am start -n org.tensorflow.demo/org.tensorflow.demo.
ClassifierActivity
$ adb shell am start -n org.tensorflow.demo/.ClassifierActivity
```

如果你不能在辦公室工作，那麼你可以透過在主機上使用鍵盤來簡化與手機的互動。當你在家工作時，它非常方便。人們喜歡的另一種方式是使用 adb port forward，你可以在本機編譯 APK 並安裝在遠端手機上。筆者認為第一種方式非常強大而且速度很快，當你在家工作時，甚至不需要觸控手機，是不是很酷？

4.3.1 程式的流程

現在，讀者應該能夠建置和執行這些 TensorFlow 應用程式了。現在讓我們來解釋和了解程式，以及學習如何在 Android 應用程式中使用 TensorFlow。

如表 4-1 所示是 Java 的執行步驟。

表 4-1 Java 的執行步驟

程式邏輯	程式路徑
應用邏輯	tensorflow/examples/android
TensorFlow 外部套件層	tensorflow/contrib/android
TensorFlow java 層	tensorflow/java

TensorFlow 提供了許多範例來幫助使用者了解 TensorFlow API 的用法。
讀者可以從 label_image 範例開始。它有三個版本，一個是用 Python 撰寫的，一個是用 C++ 撰寫的，一個是用 Java 撰寫的。

推理的基本程式流程如圖 4-2 所示。

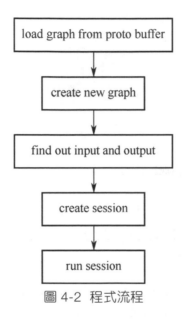

圖 4-2 程式流程

在 Python 中，讀者可能會發現 "import /" 是輸入和輸出層的字首：

```
input_name = "import/" + input_layer
output_name = "import/" + output_layer
```

這是因為 import_graph_def 加上了字首：

```
def load_graph(model_file):
  graph = tf.Graph()
  graph_def = tf.GraphDef()

  with open(model_file, "rb") as f:
```

```
    graph_def.ParseFromString(f.read())
with graph.as_default():
    **tf.import_graph_def(grap_def)**

    return graph
```

我們可以把上面反白的行寫成：

```
**tf.import_graph_def(graph_def, name="")**
```

這種修改可能更適合比對 C++ 和 Java 範例中的邏輯，因此 Python、C++ 和 Java 看起來十分類似。

4.3.2 程式的依賴性

了解程式的依賴性是了解程式的重要一環，我們可以透過執行下面的指令來列出針對 tensorflow_demo 這個應用的所有相依關係，並把相依關係輸出到 graph.dot 檔案中：

```
$ bazel query --noimplicit_deps 'deps(//tensorflow/examples/android:
tensorflow_demo)' --output graph > graph.dot
```

然後，使用 xdot 開啟檔案：

```
$ xdot graph.dot
```

在 Mac 上，先安裝 xdot，指令如下：

```
brew install xdot
```

上面的指令將建置一個相當大的依賴圖，這個圖裡包含了建置目標的所有相依關係，圖形檔案會很大，讀者可以自己產生看一下。我們可以從這個圖體會到 Blaze 強調的對每一個依賴的確定性。讀者還可以將

graphviz 的 dot 檔案轉為 jpg 或 png 影像檔,但是,在 xdot 中載入或轉換成影像檔需要很長時間。有一些技術可以進一步最佳化 graphviz 輸出檔案,但它不是本書的主題。

4.3.3 效能和程式追蹤

在示範程式中,Android Trace 已被 Trace 函數啟用,下面的程式展示了如何使用 Trace:

```
// 將輸入資料複製到 TensorFlow
Trace.beginSection("feed");
inferenceInterface.feed(inputName, floatValues, 1, inputSize, inputSize, 3);
Trace.endSection();


// 執行推理呼叫
Trace.beginSection("run");
inferenceInterface.run(outputNames, logStats);
Trace.endSection();


// 將輸出張量複製回輸出陣列
Trace.beginSection("fetch");
inferenceInterface.fetch(outputName, outputs);
Trace.endSection();
```

讓我們執行上面的程式並檢查其效能:

```
$ python /opt/Android/sdk/platform-tools/systrace/systrace.py --app org.
tensorflow.demo -t 5 -o result.html
```

在 Android Trace 執行中,我們需要指定應用程式名稱為 org.tensorflow.demo,收集追蹤資料 5 秒,輸出檔案為 result.html。

然後，載入瀏覽器就可以看到執行狀態，如圖 4-3 所示。

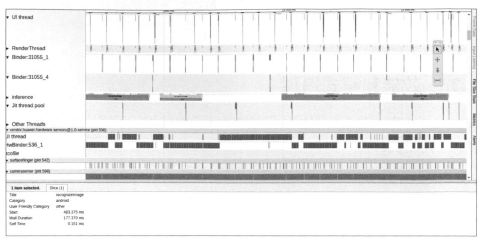

圖 4-3 執行狀態

用 TensorFlow Mobile
建置機器學習應用

本章將介紹怎樣使用 TensorFlow 開發 Android 的機器學習應用。

5.1 準備工作

為了看到更多記錄檔記錄,可以在 tensorflow/examples/android/src/org/
tensorflow/ demo/env/Logger.java 中把 Log.DEBUG 換成 Log.VERBOSE:

```
private static final String DEFAULT_TAG = "tensorflow";
private static final int DEFAULT_MIN_LOG_LEVEL = Log.DEBUG;
```

在應用中要用到模型和標記表,通常我們希望開發者能自動下載編譯,
應用也能自動讀取,我們來看一下 Bazel 是怎樣實現這個功能的。首先
我們看一下 tensorflow/ examples/android/ 下的 build 檔案:

```
android_binary( name = "tensorflow_demo",
assets = [
```

```
    "//tensorflow/examples/android/assets:asset_files",
    ":external_assets",
],
```

build 檔案裡定義了編譯目標 tensorflow_demo，以及編譯要產生的應用 tensorflow_ demo.apk。這個應用依賴於 asset 中的 :extemal assets，它在 build 檔案中的定義如下：

```
filegroup(
    name = "external_assets",
    srcs = [
        "@inception_v1//:model_files",
        "@mobile_ssd//:model_files",
        "@speech_commands//:model_files",
        "@stylize//:model_files",
    ],
)
```

filegroup 定義了一組檔案，這些檔案就是模型和標記檔案。tensorflow_ demo 這個應用會把其中定義的四個目標包含的檔案全部儲存到 APK 裡的 assert 下面。由於模型檔案佔用儲存空間較大，所以通常只儲存所需模型即可。現在看看其中的建置目標 "@inception_v1//:model_files"，它的定義可以在 workspace 檔案裡找到，workspace 檔案程式如下：

```
http_archive(
    name = "inception_v1",
    build_file = "//:models.BUILD",
    sha256 = "7efe12a8363f09bc24d7b7a450304a15655a57a7751929b2c1593a71183bb105",
    urls = [
        "http://storage.googleapis.com/download.tensorflow.org/models/
inception_v1.zip",
        "http://download.tensorflow.org/models/inception_v1.zip",
    ],
)
```

Bazel 會自動從 http://storage.googleapis.com/download.tensorflow.org/models/ inception_ v1.zip 或 http://download.tensorflow.org/models/inception_v1.zip 下載解壓檔案,並檢查 sha256 值。下載成功後,使用 "//:models.BUILD" 進行建置,實作方式程式如下:

```
filegroup(
    name = "model_files",
    srcs = glob(
        [
            "**/*",
        ],
        exclude = [
            "**/BUILD",
            "**/WORKSPACE",
            "**/LICENSE",
            "**/*.zip",
        ],
    ),
)
```

這段程式會自動忽略下載檔案中的一些檔案,例如 BUILD、WORKSPACE 等,而把其他檔案作為建置目標,這樣其他建置目標可以參照這些檔案。

舉例來說,我們可以透過下面的指令下載模型檔案:

```
$ wget "http://storage.googleapis.com/download.tensorflow.org/models/
inception_v1.zip"
$ unzip inception_v1.zip
$ ls -all
-r--r----- 1    10492 Nov 18  2015 imagenet_comp_graph_label_strings.txt
-rw-r----- 1 49937249 Jan 22  2018 inception_v1.zip
-r--r----- 1    11416 Nov 18  2015 LICENSE
-rw-r----- 1 53881635 Sep 28  2017 tensorflow_inception_graph.pb
```

獲得這些檔案並成功建置 tensorflow_demo.apk 後，我們可以執行下面的指令：

```
$ unzip bazel-bin/tensorflow/examples/android/tensorflow_demo.apk
$ ls -all asserts
-rw-rw-rw- 1        328 Jan  1   2010 BUILD.bazel
-rw-rw-rw- 1        665 Jan  1   2010 coco_labels_list.txt
-rw-rw-rw- 1    3771239 Jan  1   2010 conv_actions_frozen.pb
-rw-rw-rw- 1         60 Jan  1   2010 conv_actions_labels.txt
-rw-rw-rw- 1      10492 Jan  1   2010 imagenet_comp_graph_label_strings.txt
-rw-rw-rw- 1      11416 Jan  1   2010 LICENSE
-rw-rw-rw- 1   29083865 Jan  1   2010 ssd_mobilenet_v1_android_export.pb
-rw-rw-rw- 1     563897 Jan  1   2010 stylize_quantized.pb
-rw-rw-rw- 1   53881635 Jan  1   2010 tensorflow_inception_graph.pb
drwxr-x--- 2       4096 Jan 21  11:01 thumbnails
-rw-rw-rw- 1         28 Jan  1   2010 WORKSPACE
```

我們定義的模型檔案和標記目的檔案都被儲存在 asserts 下面了。請注意，tensorflow_ inception_graph.pb 檔案大小近 54MB，stylize_quantized.pb 檔案大小不到 600KB。模型佔用的儲存空間還是不小的。APK 的大小，不僅影響了對裝置儲存的要求，而且，使用者第一次下載要花費大量時間，這對使用者體驗也有很大影響。

5.2 影像分類（Image Classification）

影像分類是人工智慧的主要應用，我們來看一下怎樣在行動裝置上實現影像的分類。

5.2.1 應用

下面的實例主要說明 tensorflow/examples/android/src/org/tensorflow/demo 下的 Tensor- FlowImageClassifier.java 檔案。

影像分類的 Activity 是 tensorflow/examples/android/src/org/tensorflow/demo/ ClassifierActivity. java。它的定義是：

```
public class ClassifierActivity extends CameraActivity implements
OnImageAvailableListener {}
```

它 繼 承 了 CameraActivity， 並 實 現 了 OnImageAvailableListener。 CameraActivity 實現了 Android 應用的基本生命週期的功能，例如 onStart、onCreate、onStop、onDestroy 等。

另外，它實現了相機的預覽（Preview）。實現預覽的主要原因是，我們 要從相機裡取得影像的資料。在 onCreate 裡首先呼叫 setFragment，在這 段程式裡會產生 Camera- ConnectionFragment 的實例。

```
if (useCamera2API) {
  CameraConnectionFragment camera2Fragment =
      CameraConnectionFragment.newInstance(
        new CameraConnectionFragment.ConnectionCallback() {
          @Override
          public void onPreviewSizeChosen(final Size size, final int
rotation) {
            previewHeight = size.getHeight();
            previewWidth = size.getWidth();
            CameraActivity.this.onPreviewSizeChosen(size, rotation);
          }
        },
        this,
        getLayoutId(),
```

```
            getDesiredPreviewFrameSize());

  camera2Fragment.setCamera(cameraId);
  fragment = camera2Fragment;
} else {
  fragment =
      new LegacyCameraConnectionFragment(this, getLayoutId(),
getDesiredPreviewFrameSize());
}
```

tensorflow/examples/android/src/org/tensorflow/demo 下 的 CameraConnection
Fragment.Java 檔 案 實 現 了 CameraConnectionFragment，這 是 Android 的
Fragment。其關鍵的功能是由 setUpCameraOutputs 實現的，程式如下：

```
  private void setUpCameraOutputs() {
    final Activity activity = getActivity();
    final CameraManager manager = (CameraManager) activity.getSystemService
(Context.CAMERA_SERVICE);
    try {
      final CameraCharacteristics characteristics = manager.
getCameraCharacteristics(cameraId);

      final StreamConfigurationMap map =
          characteristics.get(CameraCharacteristics.SCALER_STREAM_
CONFIGURATION_MAP);

      // 使用最大尺寸進行影像抓取
      final Size largest =
          Collections.max(
              Arrays.asList(map.getOutputSizes(ImageFormat.YUV_ 420_888)),
              new CompareSizesByArea());

      sensorOrientation = characteristics.get(CameraCharacteristics.
```

```
SENSOR_ORIENTATION);

    // 預覽尺寸過大會超過相框
    // 垃圾捕捉資料
    previewSize =
        chooseOptimalSize(map.getOutputSizes(SurfaceTexture.class),
            inputSize.getWidth(),
            inputSize.getHeight());

    // TextureView 的長寬比要與我們選擇的預覽尺寸相比對
    final int orientation = getResources().getConfiguration(). orientation;
    if (orientation == Configuration.ORIENTATION_LANDSCAPE) {
      textureView.setAspectRatio(previewSize.getWidth(), previewSize.
getHeight());
    } else {
      textureView.setAspectRatio(previewSize.getHeight(), previewSize.
getWidth());
    }
  } catch (final CameraAccessException e) {
    LOGGER.e(e, "Exception!");
  } catch (final NullPointerException e) {
    // 當此程式執行的裝置不支援 Camera2API 時，拋棄 NPE
    // 此程式執行的裝置

    ErrorDialog.newInstance(getString(R.string.camera_error))
        .show(getChildFragmentManager(), FRAGMENT_DIALOG);
    throw new RuntimeException(getString(R.string.camera_error));
  }

  cameraConnectionCallback.onPreviewSizeChosen(previewSize,
sensorOrientation);
  }
```

在透過 getCameraCharacteristics(cameraId) 獲得相機的屬性後，透過與影像的解像度進行比較，呼叫 chooseOptimalSize 找到最合適的預覽 Preview 的尺寸。

```java
protected static Size chooseOptimalSize(final Size[] choices, final int
width, final int height) {
  final int minSize = Math.max(Math.min(width, height), MINIMUM_PREVIEW_SIZE);
  final Size desiredSize = new Size(width, height);

  // 收集支援的解析度，這些解析度至少與預覽圖面一樣大
  boolean exactSizeFound = false;
  final List<Size> bigEnough = new ArrayList<Size>();
  final List<Size> tooSmall = new ArrayList<Size>();
  for (final Size option : choices) {
    if (option.equals(desiredSize)) {
      // 設定尺寸，但不傳回，以便記錄剩餘尺寸
      exactSizeFound = true;
    }

    if (option.getHeight() >= minSize && option.getWidth() >= minSize) {
      bigEnough.add(option);
    } else {
      tooSmall.add(option);
    }
  }

  LOGGER.i("Desired size: " + desiredSize + ", min size: " + minSize + "x" +
minSize);
  LOGGER.i("Valid preview sizes: [" + TextUtils.join(", ", bigEnough) + "]");
  LOGGER.i("Rejected preview sizes: [" + TextUtils.join(", ", tooSmall) + "]");

  if (exactSizeFound) {
    LOGGER.i("Exact size match found.");
```

```
    return desiredSize;
  }

  // 選擇其中最小的影像
  if (bigEnough.size() > 0) {
    final Size chosenSize = Collections.min(bigEnough, new
CompareSizesByArea());
    LOGGER.i("Chosen size: " + chosenSize.getWidth() + "x" + chosenSize.
getHeight());
    return chosenSize;
  } else {
    LOGGER.e("Couldn't find any suitable preview size");
    return choices[0];
  }
}
```

如果找到完全符合的影像就傳回，否則傳回較小的影像。

到這裡我們基本上就實現了把一個相機設定好並取得其預覽的功能。由
於我們在這裡需要建置一個完整的應用，所以要透過大量程式來實現從
一個實際的應用中取得影像程式的功能。在測試或非應用的程式中可以
選用一些靜態的影像，程式也會簡單很多。

回到 CameraActivity.java 的 setFragment，下面的程式會顯示 Fragment：

```
getFragmentManager()
    .beginTransaction()
    .replace(R.id.container, fragment)
    .commit();
```

在 onResume() 裡面，主要做了兩件事，一是建立一個後台執行緒，二是
啟動相機。啟動相機就是呼叫 Android CameraManager 的 openCamera 函
數，這個應用做了一個簡單的封裝，程式如下：

```
@Override
  public void onResume() {
    super.onResume();
    startBackgroundThread();

    // 當螢幕關閉並重新開啟時，surfacetexture 為可用狀態，並且不會呼叫
"onsurfaceextureavailable"，我們可以開啟一個攝影機進行預覽
    if (textureView.isAvailable()) {
      openCamera(textureView.getWidth(), textureView.getHeight());
    } else {
      textureView.setSurfaceTextureListener(surfaceTextureListener);
    }
  }
```

建立後台執行緒。過程很簡單，這裡呼叫 Android 的 Handler 和 HandlerThread，產生並啟動一個名為 "ImageListener" 的執行緒。

```
private void startBackgroundThread() {
  backgroundThread = new HandlerThread("ImageListener");
  backgroundThread.start();
  backgroundHandler = new Handler(backgroundThread.getLooper());
}
```

啟動相機。如果相機被開啟和啟動，那麼下面的函數就會被呼叫，並啟動 createCamera- PreviewSession。

```
private final CameraDevice.StateCallback stateCallback =
    new CameraDevice.StateCallback() {
      @Override
      public void onOpened(final CameraDevice cd) {
        // 當相機開啟時，這個方法就會被呼叫，我們即可開始預覽相機

        cameraOpenCloseLock.release();
```

```
    cameraDevice = cd;
    createCameraPreviewSession();
  }
```

使用相機預覽 Preview 的應用，基本要實現兩個功能，一是設定相機的預覽大小，二是實現一個相機輸入階段 CameraCaptureSession，它的實現在

private void createCameraPreviewSession() 中。實作方式過程如下：

首先，建立一個肌理（Texture）和與其連結的 TextureView，作為影像輸出的顯示區，這樣我們就可以在裝置上看到預覽影像：

```
final SurfaceTexture texture = textureView.getSurfaceTexture();
assert texture != null;

// 我們將預設緩衝區的大小設定為所需的相機預覽大小
texture.setDefaultBufferSize(previewSize.getWidth(), previewSize.getHeight());

// 這是我們需要開始預覽的輸出曲面
final Surface surface = new Surface(texture);

// 我們用輸出曲面設定了 CaptureRequest.Builder
previewRequestBuilder = cameraDevice.createCaptureRequest(CameraDevice.
TEMPLATE_PREVIEW);
previewRequestBuilder.addTarget(surface);
```

然後，新增一個 ImageReader，並設定預覽大小和圖像資料的回呼：

```
// 為預覽畫面建立讀卡機
previewReader =
    ImageReader.newInstance(
        previewSize.getWidth(), previewSize.getHeight(), ImageFormat.
```

```
YUV_420_888, 2);

previewReader.setOnImageAvailableListener(imageListener, backgroundHandler);
previewRequestBuilder.addTarget(previewReader.getSurface());
```

最後，呼叫 cameraDevice.createCaptureSession 產生一個相機的影像抓取階段，關鍵的地方是產生一個階段的回呼，程式如下：

```
public void onConfigured(final CameraCaptureSession cameraCaptureSession) {
    // 相機已關閉
    if (null == cameraDevice) {
        return;
    }
    // 當階段準備就緒時，開始顯示預覽
captureSession = cameraCaptureSession;
    try {
        // 在相機預覽時，自動對焦是連續的
        previewRequestBuilder.set(
            CaptureRequest.CONTROL_AF_MODE,
            CaptureRequest.CONTROL_AF_MODE_CONTINUOUS_PICTURE);
        // 必要時自動啟用快閃記憶體
previewRequestBuilder.set(
            CaptureRequest.CONTROL_AE_MODE, CaptureRequest.CONTROL_AE_
MODE_ON_AUTO_FLASH);
        // 顯示相機預覽
        previewRequest = previewRequestBuilder.build();
        captureSession.setRepeatingRequest(
            previewRequest, captureCallback, backgroundHandler);
    } catch (final CameraAccessException e) {
        LOGGER.e(e, "Exception!");
    }
}
```

至此，這個應用的基本準備功能和架構都有了，可以從相機看到影像，圖像資料透過回呼函數也可以獲得。下面把應用和影像分類的模型聯繫起來。

ClassifierActivity 實現了 OnImageAvailableListener，也實現了 Camera. PreviewCallback，當新的影像被相機產生後，onImageAvailable 會被呼叫。這個函數主要做兩個工作，一是影像格式的轉換，二是呼叫 processImage() 進行影像處理。

```
@Override
public void onImageAvailable(final ImageReader reader)
```

下面是函數的定義。它的影像轉換主要呼叫 ImageUtils 類別裡的函數。

```
Trace.beginSection("imageAvailable");
final Plane[] planes = image.getPlanes();
fillBytes(planes, yuvBytes);
yRowStride = planes[0].getRowStride();
final int uvRowStride = planes[1].getRowStride();
final int uvPixelStride = planes[1].getPixelStride();

imageConverter =
    new Runnable() {
      @Override
      public void run() {
        ImageUtils.convertYUV420ToARGB8888(
            yuvBytes[0],
            yuvBytes[1],
            yuvBytes[2],
            previewWidth,
            previewHeight,
            yRowStride,
            uvRowStride,
```

```
            uvPixelStride,
            rgbBytes);
    }
};
```

然後，呼叫 processImage() 函數：

```
Trace.beginSection("imageAvailable");
Trace.endSection();
```

該函數使用 Android 的 Trace 功能，可以記錄影像處理所需要的時間。

請注意，在 ImageUtils 裡有幾個 YUV 和 RGB 轉換的函數。YUV 在硬體影像處理裡使用較多，多數相機的輸出格式也支援 YUV。RGB 是一種歷史很長的格式，它代表了紅綠藍在顏色裡的組成，比較容易了解，很多應用會使用這種格式。所以，不同顏色的轉換是必要的，例如 convertYUV420ToARGB8888，相關轉換程式如下：

```
int yp = 0;
for (int j = 0; j < height; j++) {
    int pY = yRowStride * j;
    int pUV = uvRowStride * (j >> 1);

    for (int i = 0; i < width; i++) {
        int uv_offset = pUV + (i >> 1) * uvPixelStride;

        out[yp++] = YUV2RGB(
            0xff & yData[pY + i],
            0xff & uData[uv_offset],
            0xff & vData[uv_offset]);
    }
}
```

對於每一個像素，依據下面的公式進行轉換：

```
private static int YUV2RGB(int y, int u, int v) {
  // nR = (int)(1.164 * nY + 2.018 * nU);
  // nG = (int)(1.164 * nY - 0.813 * nV - 0.391 * nU);
  // nB = (int)(1.164 * nY + 1.596 * nV);
}
```

細心的讀者一定會注意到，這個函數很簡單，是簡單的 for 循環和加乘法的組合，也一定會消耗很多時間，如果是 512×512 的影像，會重複很多簡單計算。在影像處理中，透過最佳化函數來提高性能的方法有幾種，可以用機器原生語言 C 或組合語言來實現，也可以使用機器上的硬體加速來實現。

實際上，在實測中我們也發現了這個問題，但是這是一個示範應用，並不多佔機器效能，因此就採用了現在這個方案。

這個應用會獲得兩個回呼，一是相機本身產生的影像，二是預覽 Preview 的影像，我們會根據應用採用不同的處理方式。

為了實現影像分類，在 ClassifierActivity 裡多載了 processImage，並呼叫分類模型，輸出分類的結果：

```
@Override
protected void processImage() {
  rgbFrameBitmap.setPixels(getRgbBytes(), 0, previewWidth, 0, 0,
previewWidth, previewHeight);
  final Canvas canvas = new Canvas(croppedBitmap);
  canvas.drawBitmap(rgbFrameBitmap, frameToCropTransform, null);

  // 檢查目前的 TF 輸入
  if (SAVE_PREVIEW_BITMAP) {
    ImageUtils.saveBitmap(croppedBitmap);
```

```
    }
    runInBackground(
        new Runnable() {
          @Override
          public void run() {
            final long startTime = SystemClock.uptimeMillis();
            final List<Classifier.Recognition> results = classifier.
recognizeImage(croppedBitmap);
            lastProcessingTimeMs = SystemClock.uptimeMillis() - startTime;
            LOGGER.i("Detect: %s", results);
            cropCopyBitmap = Bitmap.createBitmap(croppedBitmap);
            if (resultsView == null) {
              resultsView = (ResultsView) findViewById(R.id.results);
            }
            resultsView.setResults(results);
            requestRender();
            readyForNextImage();
          }
        });
    }
```

classifier.recognizeImage 實現了影像分類，它的輸入是一個 BitMap，然
後傳回一個分類結果的佇列：

```
final List<Classifier.Recognition> results = classifier.recognizeImage
(croppedBitmap);
```

由於模型的預測需要耗費運算資源和時間，這個函數一定要執行在非主
執行緒上，這裡使用了 Runnable。分類的結果是一個類別 Recognition，
包含了分類的結果，它的定義如下：

```
public class Recognition {
  // 已識別內容的唯一識別碼。特定於類別，而非物件
```

```java
private final String id;
// 顯示識別名稱
private final String title;
// 一個可排序的分數,表示相對於其他人的認可度,該認可度值越高越好
private final Float confidence;
// 源影像中可用於識別物件位置的可選位置
private RectF location;
```

title 是被分類物體的名稱,confidence 是傳回的與分類物體相對應的信心
數,這個數值越高越好。

下面是影像分類的實現過程,它由幾部分組成。

第一步,先把影像的 RGB 數值轉為浮點值:

```java
bitmap.getPixels(intValues, 0, bitmap.getWidth(), 0, 0, bitmap.getWidth(),
bitmap.getHeight());
for (int i = 0; i < intValues.length; ++i) {
  final int val = intValues[i];
  floatValues[i * 3 + 0] = (((val >> 16) & 0xFF) - imageMean) / imageStd;
  floatValues[i * 3 + 1] = (((val >> 8) & 0xFF) - imageMean) / imageStd;
  floatValues[i * 3 + 2] = ((val & 0xFF) - imageMean) / imageStd;
}
```

第二步,把浮點值傳入模型中進行預測:

```java
inferenceInterface.feed(inputName, floatValues, 1, inputSize, inputSize, 3);
```

第三步,執行模型的階段:

```java
inferenceInterface.run(outputNames, logStats);
```

第四步,取出模型的預測結果:

```java
inferenceInterface.fetch(outputName, outputs);
```

以上幾步，除了第一步，基本和用 Python 實現預測的步驟是一樣的，只不過是用 Java 實現的。

下一步是對輸出的結果進行比較，最後輸出的處理比較簡單，一是比較信心值，只顯示幾個比較有意義的結果，二是省去其他信心值比較低的結果。實作方式程式如下：

```
// 找到最佳分類
PriorityQueue<Recognition> pq =
    new PriorityQueue<Recognition>(
        3,
        new Comparator<Recognition>() {
          @Override
          public int compare(Recognition lhs, Recognition rhs) {
              // 故意顛倒，以在佇列的最前面建立高度的信心
return Float.compare(rhs.getConfidence(), lhs.getConfidence());
          }
        });
for (int i = 0; i < outputs.length; ++i) {
  if (outputs[i] > THRESHOLD) {
    pq.add(
        new Recognition(
            "" + i, labels.size() > i ? labels.get(i) : "unknown",
outputs[i], null));
  }
}
```

在程式中對每一個步驟都使用了 Trace，我們可以使用工具更進一步地了解每一步消耗的時間，並根據結果做最佳化。

```
Trace.beginSection("feed");
Trace.endSection();
```

5.2.2 模型

這個應用的模型是 tensorflow_inception_graph.pb，由下面的程式定義。

```
private static final String MODEL_FILE = "file:///android_asset/tensorflow_
inception_graph.pb";
  private static final String LABEL_FILE =
      "file:///android_asset/imagenet_comp_graph_label_strings.txt";
```

模型的原始檔案程式如下：

```
http_archive(
    name = "inception_v1",
    build_file = "//:models.BUILD",
    sha256 = "7efe12a8363f09bc24d7b7a450304a15655a57a7751929b2c1593a71183
bb105",
    urls = [
        "http://storage.googleapis.com/download.tensorflow.org/models/
inception_v1.zip",
        "http://download.tensorflow.org/models/inception_v1.zip",
    ],
)
```

我們可以下載並解壓這個模型，獲得 tensorflow_inception_graph.pb，然後執行以下程式：

```
$ python tensorflow/python/tools/import_pb_to_tensorboard.py --model_dir
tensorflow_inception_graph.pb --log_dir /tmp/log
$ tensorboard --logdir /tmp/log
```

開啟瀏覽器，輸入位址 http://localhost:6006，可以看到 Inception 模型圖，如圖 5-1 所示。

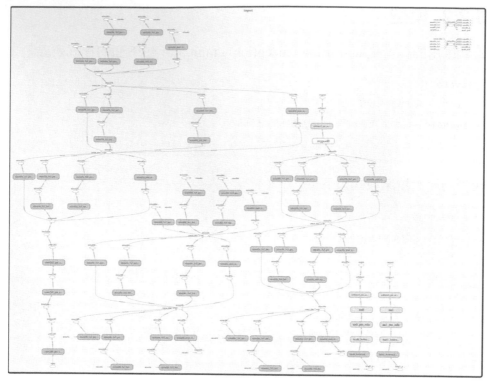

圖 5-1 Inception 模型圖

5.3 物體檢測（Object Detection）

5.3.1 應用

DetectorActivity.java 實現了使用機器學習模型來進行物體檢測。這個應用只是對相機的影像進行處理，所以只繼承了 OnImageAvailableListener：

```
public class DetectorActivity extends CameraActivity implements
OnImageAvailableListener
```

這個實例提供了三種模式，它們的定義如下：

```
private enum DetectorMode {
  TF_OD_API, MULTIBOX, YOLO;
}
```

預設值是 YOLO，讀者可以改變這個值，重新編譯執行這個應用，比較三個模型的差異。執行這三個模型的方式是基本一致的。

但是，對於信心（confidence）值，不同的模型會有不同處理，應用裡分別定義了對應的數值：

```
// 追蹤檢測的最小可靠度
private static final float MINIMUM_CONFIDENCE_TF_OD_API = 0.6f;
private static final float MINIMUM_CONFIDENCE_MULTIBOX = 0.1f;
private static final float MINIMUM_CONFIDENCE_YOLO = 0.25f;
```

另外，模型對於輸入影像有不同的要求，它們輸入的變數名稱也不同。而且，應用把模型用不同的檔案名稱儲存起來，下面是各自的定義。這些常數其實可以封裝起來，由 MODE 來決定，程式看起來會更簡單一些。

```
private static final int MB_INPUT_SIZE = 224;
private static final int MB_IMAGE_MEAN = 128;
private static final float MB_IMAGE_STD = 128;
private static final String MB_INPUT_NAME = "ResizeBilinear";
private static final String MB_OUTPUT_LOCATIONS_NAME = "output_ locations/
Reshape";
private static final String MB_OUTPUT_SCORES_NAME = "output_ scores/Reshape";
private static final String MB_MODEL_FILE = "file:///android_asset/
multibox_model.pb";
private static final String MB_LOCATION_FILE =
    "file:///android_asset/multibox_location_priors.txt";

private static final int TF_OD_API_INPUT_SIZE = 300;
```

```
private static final String TF_OD_API_MODEL_FILE =
    "file:///android_asset/ssd_mobilenet_v1_android_export.pb";
private static final String TF_OD_API_LABELS_FILE = "file: ///android_asset/
coco_labels_list.txt";

private static final String YOLO_MODEL_FILE = "file:///android_asset/graph-
tiny-yolo-voc.pb";
private static final int YOLO_INPUT_SIZE = 416;
private static final String YOLO_INPUT_NAME = "input";
private static final String YOLO_OUTPUT_NAMES = "output";
private static final int YOLO_BLOCK_SIZE = 32;
```

以下程式產生了 MultiBoxTracker。MultiBoxTracker 負責追蹤檢測物體
並把物體的外框（Box）表示出來。DetectorActivity 會把物體檢測的結
果傳進來，並由 MultiBoxTracker 顯示到裝置上。

```
tracker = new MultiBoxTracker(this);
```

下面的程式根據 MODE 產生了檢測器的實例。這三個檢測器都實現了介
面 public interface Classifier，所以都被封裝了起來，使用者也不用關心
它們的實現細節。

```
if (MODE == DetectorMode.YOLO) {
   detector =
       TensorFlowYoloDetector.create(
           getAssets(),
           YOLO_MODEL_FILE,
           YOLO_INPUT_SIZE,
           YOLO_INPUT_NAME,
           YOLO_OUTPUT_NAMES,
           YOLO_BLOCK_SIZE);
   cropSize = YOLO_INPUT_SIZE;
} else if (MODE == DetectorMode.MULTIBOX) {
   detector =
```

```
        TensorFlowMultiBoxDetector.create(
            getAssets(),
            MB_MODEL_FILE,
            MB_LOCATION_FILE,
            MB_IMAGE_MEAN,
            MB_IMAGE_STD,
            MB_INPUT_NAME,
            MB_OUTPUT_LOCATIONS_NAME,
            MB_OUTPUT_SCORES_NAME);
    cropSize = MB_INPUT_SIZE;
  } else {
    try {
      detector = TensorFlowObjectDetectionAPIModel.create(
          getAssets(), TF_OD_API_MODEL_FILE, TF_OD_API_LABELS_FILE,
  TF_OD_ API_INPUT_SIZE);
      cropSize = TF_OD_API_INPUT_SIZE;
  }
```

在影像識別（Image Classification）的實例中，我們關心被識別的物體是什麼，在這個應用裡，我們也關心物體的位置，所以在介面 Classifier 裡還定義了一個類型為 android.graphics.RectF 的位置變數：

```
// 源影像中可用於識別物件位置的可選位置
private RectF location;
```

Tracker 和 detector 完成了這個應用的主要功能。Detector 負責呼叫模型，傳回檢測到的物體名稱和位置。Tracker 負責把檢測到的物體顯示到顯示器上。在這個應用裡，為了顯示和追蹤物體會花費大量的程式，包含 MultiBoxTracker.java、ObjectTracker.java 和位於 tensorflow/examples/android/jni/object_tracking 下的大量 C++ 程式。

模型會傳回檢測到的物體名稱和位置。但是，傳回的物體名稱和位置不是連續的，如果我們只是按照模型傳回的數值進行顯示是絕對不夠的，在現實中一般物體的移動速度是有限的，移動的位置是有連結性的。按照這個原理，我們可以透過程式實現簡單的物體追蹤。在機器學習的應用裡，實際上撰寫大量的程式還是為了能實現應用的邏輯功能。

這個應用的主要功能是在 processImage() 裡實現的，分別呼叫了 Tracker 和 Detector 的函數。一個簡化的實現過程大概有三個步驟：

```
@Override
 protected void processImage() {
    // 第一步，清除已檢測的物體，準備下一次顯示
    ++timestamp;
    tracker.onFrame(
        previewWidth,
        previewHeight,
        getLuminanceStride(),
        sensorOrientation,
        originalLuminance,
        timestamp);
    trackingOverlay.postInvalidate();

    runInBackground(
        new Runnable() {
          @Override
          public void run() {
            // 第二步，呼叫模型，取得被識別物體的佇列，包含物體的名稱和位置
            LOGGER.i("Running detection on image " + currTimestamp);
            final long startTime = SystemClock.uptimeMillis();
            final List<Classifier.Recognition> results = detector.
recognizeImage(croppedBitmap);
            lastProcessingTimeMs = SystemClock.uptimeMillis() - startTime;

            // 第三步，把已識別的物體顯示出來
            tracker.trackResults(mappedRecognitions, luminanceCopy,
currTimestamp);
```

```
            trackingOverlay.postInvalidate();

            requestRender();
            computingDetection = false;
          }
        });
}
```

我們先看一下 detector 的實現。上面提到了，這個應用支援三個檢測
器，三個檢測器可實現介面 classifier，它們的使用方法比較類似，都實
現了 recognizeImage，但也有些不同。以下是三個檢測器檢測的實作方
式方式，讀者可以看看它們的不同。

❑ TensorFlowMultiBoxDetector.java

```
Trace.beginSection("preprocessBitmap");
// 對圖像資料進行前置處理，轉化為標準化浮點數
bitmap.getPixels(intValues, 0, bitmap.getWidth(), 0, 0, bitmap.getWidth(),
bitmap.getHeight());

for (int i = 0; i < intValues.length; ++i) {
  floatValues[i * 3 + 0] = (((intValues[i] >> 16) & 0xFF) - imageMean) /
imageStd;
  floatValues[i * 3 + 1] = (((intValues[i] >> 8) & 0xFF) - imageMean) /
imageStd;
  floatValues[i * 3 + 2] = ((intValues[i] & 0xFF) - imageMean) / imageStd;
}
Trace.endSection();
// 點陣圖前置處理

// 將輸入資料複製到 TensorFlow 中
Trace.beginSection("feed");
inferenceInterface.feed(inputName, floatValues, 1, inputSize, inputSize, 3);
```

```
Trace.endSection();

// 執行推理呼叫
Trace.beginSection("run");
inferenceInterface.run(outputNames, logStats);
Trace.endSection();

// 將輸出張量複製回輸出陣列
Trace.beginSection("fetch");
final float[] outputScoresEncoding = new float[numLocations];
final float[] outputLocationsEncoding = new float[numLocations * 4];
inferenceInterface.fetch(outputNames[0], outputLocationsEncoding);
inferenceInterface.fetch(outputNames[1], outputScoresEncoding);
Trace.endSection();
```

❑ TensorFlowObjectDetectionAPIModel.java：

```
Trace.beginSection("preprocessBitmap");
// 前置處理圖像資料，從 0x00rrggbb 格式的 int 中分析 r、g 和 b 位元組
bitmap.getPixels(intValues, 0, bitmap.getWidth(), 0, 0, bitmap.getWidth(),
bitmap.getHeight());

for (int i = 0; i < intValues.length; ++i) {
  byteValues[i * 3 + 2] = (byte) (intValues[i] & 0xFF);
  byteValues[i * 3 + 1] = (byte) ((intValues[i] >> 8) & 0xFF);
  byteValues[i * 3 + 0] = (byte) ((intValues[i] >> 16) & 0xFF);
}
Trace.endSection();
// 點陣圖前置處理
// 將輸入資料複製到 TensorFlow 中
Trace.beginSection("feed");
inferenceInterface.feed(inputName, byteValues, 1, inputSize, inputSize, 3);
Trace.endSection();
```

```
// 執行推理呼叫
Trace.beginSection("run");
inferenceInterface.run(outputNames, logStats);
Trace.endSection();
```

❑ TensorFlowYoloDetector.java：

```
Trace.beginSection("preprocessBitmap");
// 對圖像資料進行前置處理，轉化為標準化浮點數
bitmap.getPixels(intValues, 0, bitmap.getWidth(), 0, 0, bitmap.getWidth(),
bitmap.getHeight());

for (int i = 0; i < intValues.length; ++i) {
  floatValues[i * 3 + 0] = ((intValues[i] >> 16) & 0xFF) / 255.0f;
  floatValues[i * 3 + 1] = ((intValues[i] >> 8) & 0xFF) / 255.0f;
  floatValues[i * 3 + 2] = (intValues[i] & 0xFF) / 255.0f;
}
Trace.endSection(); // 點陣圖前置處理
// 將輸入資料複製到 TensorFlow 中
Trace.beginSection("feed");
inferenceInterface.feed(inputName, floatValues, 1, inputSize, inputSize, 3);
Trace.endSection();

timer.endSplit("ready for inference");

// 執行推理呼叫
Trace.beginSection("run");
inferenceInterface.run(outputNames, logStats);
Trace.endSection();
```

在 TensorFlowObjectDetectionAPIModel 裡，模型的資料登錄類型是 byte，而 TensorFlowYoloDetector 和 TensorFlowMultiBoxDetector 的輸入類型是 float，一個是定點數，另一個是浮點數。

5.3.2 模型

這個應用用到了三個模型，其中 inception 模型在上面已經介紹了。現在來看看另外兩個，一個是 Mobile Net。

```
http_archive(
    name = "mobile_ssd",
    build_file = "//:models.BUILD",
    sha256 = "bddd81ea5c80a97adfac1c9f770e6f55cbafd7cce4d3bbe15fbeb041e6b8
f3e8",
    urls = [
        "http://storage.googleapis.com/download.tensorflow.org/models/
object_detection/ssd_mobilenet_v1_android_export.zip",
        "http://download.tensorflow.org/models/object_detection/
ssd_mobilenet_v1_android_export.zip",
    ],
)
```

使用前面介紹過的過程，模型結構的視圖如圖 5-2 所示。

另外一個是 Mobile Multibox 模型，程式如下：

```
http_archive(
    name = "mobile_multibox",
    build_file = "//:models.BUILD",
    sha256 = "859edcddf84dddb974c36c36cfc1f74555148e9c9213dedacf1d6b613ad5
2b96",
    urls = [
        "http://storage.googleapis.com/download.tensorflow.org/models/
mobile_multibox_v1a.zip",
        "http://download.tensorflow.org/models/mobile_multibox_v1a.zip",
    ],
)
```

模型的視覺化視圖結構如圖 5-3 所示。

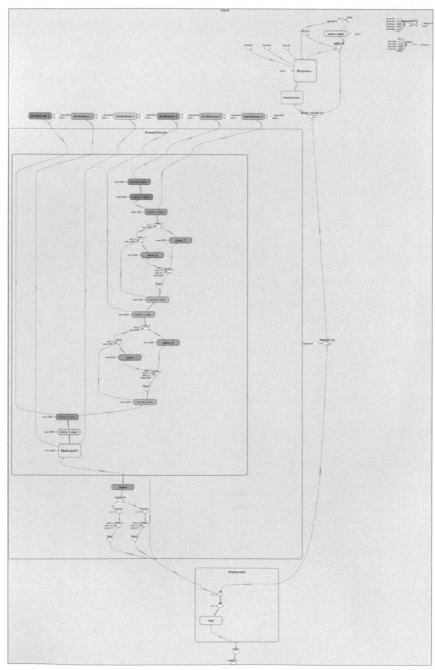

圖 5-2　SSD Mobile Net 模型圖

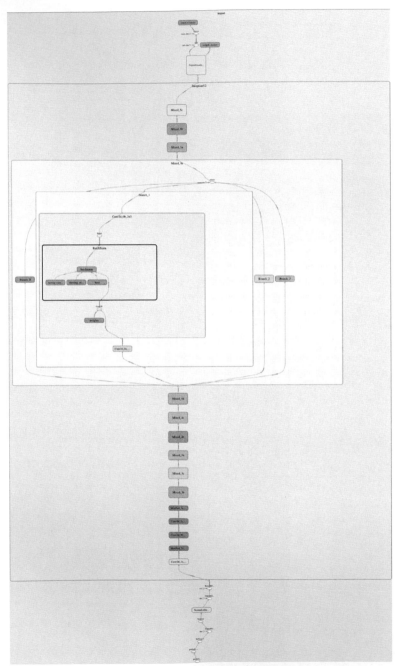

圖 5-3　Mobile Multibox 結構圖

5.4 時尚繪製（Stylization）

5.4.1 應用

時尚繪製的應用是由 StylizeActivity.java 實現的。主要的功能也是由
processImage() 呼叫 stylizeImage() 來實現的。

```
++frameNum;
bitmap.getPixels(intValues, 0, bitmap.getWidth(), 0, 0, bitmap.getWidth(),
bitmap.getHeight());


for (int i = 0; i < intValues.length; ++i) {
  final int val = intValues[i];
  floatValues[i * 3] = ((val >> 16) & 0xFF) / 255.0f;
  floatValues[i * 3 + 1] = ((val >> 8) & 0xFF) / 255.0f;
  floatValues[i * 3 + 2] = (val & 0xFF) / 255.0f;
}


// 將輸入資料複製到 TensorFlow
LOGGER.i("Width: %s , Height: %s", bitmap.getWidth(), bitmap.getHeight());
inferenceInterface.feed(
    INPUT_NODE, floatValues, 1, bitmap.getWidth(), bitmap.getHeight(), 3);
inferenceInterface.feed(STYLE_NODE, styleVals, NUM_STYLES);


inferenceInterface.run(new String[] {OUTPUT_NODE}, isDebug());
inferenceInterface.fetch(OUTPUT_NODE, floatValues);


for (int i = 0; i < intValues.length; ++i) {
  intValues[i] =
      0xFF000000
          | (((int) (floatValues[i * 3] * 255)) << 16)
          | (((int) (floatValues[i * 3 + 1] * 255)) << 8)
```

```
            | ((int) (floatValues[i * 3 + 2] * 255));
}

bitmap.setPixels(intValues, 0, bitmap.getWidth(), 0, 0, bitmap.getWidth(),
bitmap.getHeight());
```

實現的過程也相對簡單，透過呼叫 inferenceInterface 的 feed、run、fetch 後獲得像素的浮點值，然後轉換成對應的 RGB 的定點數值即可。

5.4.2 模型

時尚模型原始檔案的定義如下：

```
http_archive(
    name = "stylize",
    build_file = "//:models.BUILD",
    sha256 = "3d374a730aef330424a356a8d4f04d8a54277c425e274ecb7d9c83aa912c6bfa",
    urls = [
        "http://storage.googleapis.com/download.tensorflow.org/models/
stylize_v1.zip",
        "http://download.tensorflow.org/models/stylize_v1.zip",
    ],
)
```

我們把模型檔案下載下來，按照上面的實例，轉換成一個視覺化的圖形，程式如下：

```
$ python tensorflow/python/tools/import_pb_to_tensorboard.py --model_dir
stylize_quantized.pb --log_dir /tmp/log
```

被細化的 Style 模型如圖 5-4 所示。

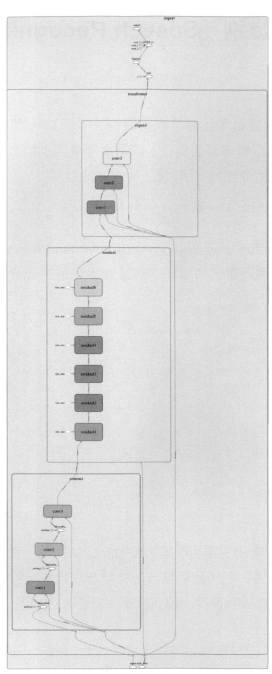

圖 5-4 被細化的 Style 模型圖

5.5 聲音識別（Speech Recognization）

5.5.1 應用

以上的幾個實例都是和影像處理有關的，本節這個實例是和聲音有關的，也是典型的機器學習的實例。這個應用會識別聲音，並把指令顯示出來。由於要使用麥克風，所以要把許可權加上：

```
<uses-permission android:name="android.permission.RECORD_AUDIO" />
```

這個應用由 tensorflow/examples/android/src/org/tensorflow/demo/SpeechActivity. java 實現。當應用啟動的時候，同時啟動麥克風的錄音和聲音識別，程式的實現方法如下：

```
protected void onCreate(Bundle savedInstanceState) {
  // 載入 TensorFlow 模型
  inferenceInterface = new TensorFlowInferenceInterface(getAssets(),
MODEL_FILENAME);

  // 啟動錄製並識別執行緒
  requestMicrophonePermission();
  startRecording();
  startRecognition();
}
```

startRecording 實現的主要功能是，啟動一個執行緒。在這個執行緒裡，首先設定錄音裝置，然後啟動錄音。在錄音開始後，把資料存到一個快取陣列裡，供聲音識別使用。程式的簡單實現方法如下：

```
private void record() {
  android.os.Process.setThreadPriority(android.os.Process.THREAD_
PRIORITY_AUDIO);
```

```
// 預估裝置需要的緩衝區大小
int bufferSize =
    AudioRecord.getMinBufferSize(
        SAMPLE_RATE, AudioFormat.CHANNEL_IN_MONO, AudioFormat.ENCODING_
PCM_16BIT);
    if (bufferSize == AudioRecord.ERROR || bufferSize == AudioRecord. ERROR_
BAD_VALUE) {
    bufferSize = SAMPLE_RATE * 2;
    }
    short[] audioBuffer = new short[bufferSize / 2];

    AudioRecord record =
        new AudioRecord(
            MediaRecorder.AudioSource.DEFAULT,
            SAMPLE_RATE,
            AudioFormat.CHANNEL_IN_MONO,
            AudioFormat.ENCODING_PCM_16BIT,
            bufferSize);

    record.startRecording();

    // 循環並收集音訊資料並將其複製到循環緩衝區
while (shouldContinue) {
    int numberRead = record.read(audioBuffer, 0, audioBuffer.length);
    int maxLength = recordingBuffer.length;
    int newRecordingOffset = recordingOffset + numberRead;
    int secondCopyLength = Math.max(0, newRecordingOffset - maxLength);
    int firstCopyLength = numberRead - secondCopyLength;
    // 儲存所有資料，以便識別執行緒存取
    // 執行緒將從這個緩衝區被複製到自己的緩衝區中，這個過程是加鎖的
    recordingBufferLock.lock();
    try {
```

```
        System.arraycopy(audioBuffer, 0, recordingBuffer, recordingOffset,
firstCopyLength);
        System.arraycopy(audioBuffer, firstCopyLength, recordingBuffer, 0,
secondCopyLength);
        recordingOffset = newRecordingOffset % maxLength;
    } finally {
        recordingBufferLock.unlock();
    }
  }
}
```

注意，程式設定的聲音格式是 SAMPLE_RATE = 16000，聲音取樣的頻率是 16k，採用的聲道是 CHANNEL_IN_MONO，聲音樣本的資料格式是 16bit 和 AudioFormat. ENCODING_PCM_16BIT。聲音識別主要是由 startRecognition() 和 recognize() 實現的，也是在另外一個執行緒上實現的。實現的主要程式如下：

```
// 輸入介於 -1.0f 和 1.0f 之間的浮點值
  for (int i = 0; i < RECORDING_LENGTH; ++i) {
    floatInputBuffer[i] = inputBuffer[i] / 32767.0f;
  }

  // 執行模型
  inferenceInterface.feed(SAMPLE_RATE_NAME, sampleRateList);
  inferenceInterface.feed(INPUT_DATA_NAME, floatInputBuffer, RECORDING_
LENGTH, 1);
  inferenceInterface.run(outputScoresNames);
  inferenceInterface.fetch(OUTPUT_SCORES_NAME, outputScores);
```

首先，要做資料類型的轉換，把 16 位元的定點數轉為浮點數。然後，還是在呼叫 feed、run 和 fetch 後，把預測的結果轉為指令顯示出來。

5.5.2 模型

時尚模型原始檔案的定義如下：

```
http_archive(
    name = "speech_commands",
    build_file = "//:models.BUILD",
    sha256 = "c3ec4fea3158eb111f1d932336351edfe8bd515bb6e87aad4f25dbad0a600d0c",
    urls = [
        "http://storage.googleapis.com/download.tensorflow.org/models/
speech_commands_v0.01.zip",
        "http://download.tensorflow.org/models/speech_commands_ v0.01.zip",
    ],
)
```

我們把模型檔案下載下來，按照上面的實例，轉換成一個視覺化的圖形，程式如下：

```
$ python tensorflow/python/tools/import_pb_to_tensorboard.py --model_dir
conv_actions_frozen.pb --log_dir /tmp/log
```

模型的視圖如圖 5-5 所示。

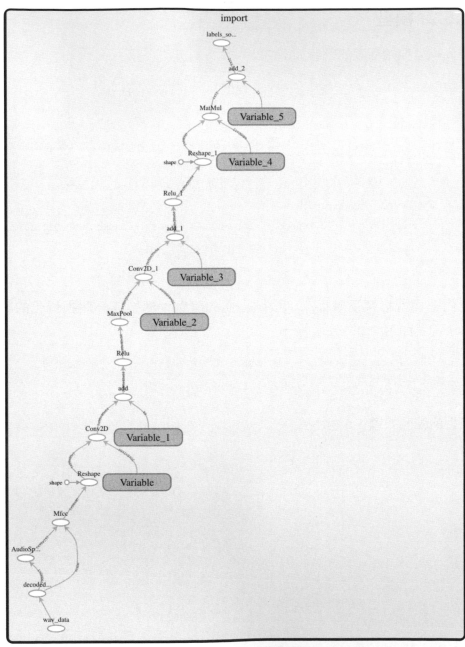

圖 5-5 Speech 模型圖

TensorFlow Lite 的架構

本章將介紹 TensorFlow Lite 的架構。TensorFlow Lite 的架構如圖 6-1 所示。

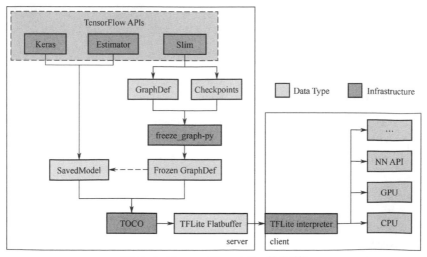

圖 6-1　TensorFlow Lite 的架構

下面介紹 TensorFlow 的主要技術特點和開發者要了解的基礎。

6.1 模型格式

首先需要了解 TensorFlow 的資料格式和它們的技術特點。

6.1.1 Protocol Buffers

Google Protocol Buffers（簡稱 Protobuf）是 Google 公司內部的混合語言資料標準，Protobuf 是一種輕便、高效的結構化資料儲存格式，可以用於結構化資料序列化或序列化。它很適合做資料儲存或 RPC 資料交換格式，可用於通訊協定、資料儲存等領域的與語言無關、平台無關、可擴充的序列化結構資料格式。它提供了包含 C++、Java、Python 在內的多種語言的 API。

Protobuf 幾乎可以稱為 Google 資料格式的基礎，是每一個 Google 工程師必須掌握的技能，而且它也在不斷地進化當中。有意思的是，Google 的團隊還在不斷改善它的效能，可以想像在資料中心裡，如果常用到這種資料結構，那麼它的效能被加強即使是千分之一、萬分之一，對整個系統性能的加強也是非常明顯的。

Protobuf 有 Version 2 和 Version 3。然而，不能說 Version 3 一定比 Version 2 好，現在有很多程式還在使用 Version 2。但是，在 TensorFlow 裡用到的都是 Version 3：

```
syntax = "proto3";
```

我們看一下 TensorFlow 的幾個 Protobuf，來了解一下它的定義，也了解一下 TensorFlow 裡的一些基本概念。

```
// 描述機器學習模型訓練或推理輸入資料範例的協定訊息
```

```
syntax = "proto3";


import "third_party/tensorflow/core/example/feature.proto";
option cc_enable_arenas = true;
option java_outer_classname = "ExampleProtos";
option java_multiple_files = true;
option java_package = "org.tensorflow.example";
// 使用 copybara 在外部增加 go 套件
package tensorflow;


message Example {
  Features features = 1;
};
```

這段程式定義了 Protobuf 本身的套件名稱 TensorFlow 和它對應 Java 套件的名稱 org.tensorflow.example。Protobuf 定義了 Example，只包含一個成員 features，它的定義在檔案 hird_party/tensorflow/core/example/feature.proto 中，程式如下：

```
syntax = "proto3";
option cc_enable_arenas = true;
option java_outer_classname = "FeatureProtos";
option java_multiple_files = true;
option java_package = "org.tensorflow.example";
// 使用 copybara 在外部增加 go 套件
package tensorflow;


// 儲存重複基本值的容器
message BytesList {
  repeated bytes value = 1;
}
message FloatList {
  repeated float value = 1 [packed = true];
```

```
}
message Int64List {
  repeated int64 value = 1 [packed = true];
}


// 非順序資料容器
message Feature {
  // 每個功能都可以是一種
  oneof kind {
    BytesList bytes_list = 1;
    FloatList float_list = 2;
    Int64List int64_list = 3;
  }
};


message Features {
  // 從要素名稱對映到要素
  map<string, Feature> feature = 1;
};
```

Features 就是字串到 Feature 的對映,而每個 Feature 只能是下面三個中的,ByteList、FloatList 或 Int64List。

另外,我們可以看到在 Protobuf 裡有大量的註釋,這是一個好的工作習慣,同時也能幫助我們了解這些定義的用法,很多文件就是從這些註釋裡直接產生的,例如:

```
// An Example is a mostly-normalized data format for storing data for
// training and inference.  It contains a key-value store (features); where
// each key (string) maps to a Feature message (which is oneof packed
BytesList,
// FloatList, or Int64List).  This flexible and compact format allows the
// storage of large amounts of typed data, but requires that the data shape
```

```
// and use be determined by the configuration files and parsers that are
used to
// read and write this format.  That is, the Example is mostly *not* a
// self-describing format.  In TensorFlow, Examples are read in row-major
// format, so any configuration that describes data with rank-2 or above
// should keep this in mind.  For example, to store an M x N matrix of Bytes,
// the BytesList must contain M*N bytes, with M rows of N contiguous values
// each.  That is, the BytesList value must store the matrix as:
```

Example 是用於樣本訓練和推理的儲存資料的規範化資料格式。它包含一個鍵值儲存功能，每個鍵（字串）對映到 Feature 的結構。Feature 是一個包裝的 BytesList、FloatList 或 Int64 的鏈結串列。這種靈活緊湊的格式可以儲存大量同類型的資料，但要求資料的形狀和使用由用於讀取和寫入此格式的設定檔和解析器來確定。

也就是說，樣本不是一種自描述格式。在 TensorFlow 中，樣本以行格式的方式讀取，因此在設定維度為 2 或更高維度數據的時候要注意這一點。舉例來說，為了儲存位元組的 $M \times N$ 矩陣，BytesList 必須包含 $M \times N$ 個位元組，每個 M 行具有 N 個連續值。

另一個實例是 Graph：

```
import "third_party/tensorflow/core/framework/node_def.proto";
import "third_party/tensorflow/core/framework/function.proto";
import "third_party/tensorflow/core/framework/versions.proto";

// 展示操作圖
message GraphDef {
  repeated NodeDef node = 1;

  VersionDef versions = 4;
```

```
  int32 version = 3 [deprecated = true];

  FunctionDefLibrary library = 2;
};
```

圖（Graph）是 TensorFlow 裡的重要的概念，從上面的定義裡我們可以
看到，所謂的圖就是一系列節點（Node）的集合，請注意裡面的兩個與
version 有關的成員：

```
  int32 version = 3 [deprecated = true];
```

這只是一個 32 位元整數型成員，而且它已經不能再被使用了
（Deprecated），取代它的是 VersionDef，它的編號是 4，而 verison 的編
號是 3。透過這種方式，Protobuf 可以極佳地解決向後相容的問題。程式
如下：

```
  VersionDef versions = 4;
```

Protobuf 還被廣泛用於定義 RPC，我們在這裡就不多描述了。

6.1.2 FlatBuffers

FlatBuffers 是一個高效的跨平台序列化函數庫，適用於 C++、C＃、
C、Go、Java、JavaScript、Lobster、Lua、TypeScript、PHP、Python 和
Rust。它最初是在 Google 建立的，用於遊戲開發及其他效能相關的關鍵
應用程式。

FlatBuffers 裡的兩個關鍵概念是 Schema 和 Table。

Table（表）的定義：Table 是 FlatBuffers 的基礎，因為格式演變對大多
數序列化應用程式而言非常重要。一般來說處理格式更改是可以在大多

數序列化解決方案的解析過程中透明地完成的，但是 FlatBuffers 在存取之前是不會被解析的。表的存取是透過使用額外的間接存取欄位來實現的，實際地講是透過 vtable 來解決這個問題的。

每個表都帶有一個 vtable（可以在具有相同版面配置的多個表之間共用），並包含儲存此特定類型的 vtable 實例的欄位資訊。vtable 也可能表示該欄位不存在（因為此 FlatBuffers 是使用較舊版本的軟體撰寫的，該實例不需要該資訊或被視為已棄用），在這種情況下傳回預設值。

表擁有很低的記憶體負擔（因為 vtable 很小並且共用）和存取成本（額外的間接），但它提供了很大的靈活性。表甚至可以比相等結構花費更少的記憶體，因為當欄位等於它們的預設值時就不需要儲存它們。FlatBuffers 還提供了「裸」結構，它不提供向前／向後相容性，但儲存的負擔可以更小（對於不太可能改變的非常小的物件有用，例如座標對或 RGBA 顏色）。

FlatBuffers 的模式與現有的 Protobuf 的模式十分類似，對熟悉 C 語言的讀者來說，FlatBuffers 的定義通常應該是易讀的。和 Protobuf 相比，FlatBuffers 透過以下方式改進 .proto 檔案提供的功能：

- 對於棄用欄位的情況，現在手動進行欄位 ID 分配。
- 區分表格和結構。所有表欄位都是可選的，所有結構欄位都是必需的。
- 使用原生的向量類型而非重複的。
- 擁有本機聯合類型 union 而非使用一系列可選欄位。
- 能夠為所有純量定義預設值，而不必在每次存取時處理其可選性。
- 一種可以統一處理模式和資料定義（JSON 相容）的解析器。

下面我們看一下，在 TensorFlow Lite 裡 schema.fbs 是如何使用 FlatBuffers 的：

```
namespace tflite;

//   對應版本
file_identifier "TFL3";
// 任何已寫入檔案的副檔名
file_extension "tflite";


// 重要提示:必須在尾端增加表、列舉和聯合的所有新成員,以確保向後的相容性


// 儲存在張量中的資料類型
enum TensorType : byte {
  FLOAT32 = 0,
  FLOAT16 = 1,
  INT32 = 2,
  UINT8 = 3,
  INT64 = 4,
  STRING = 5,
  BOOL = 6,
  INT16 = 7,
  COMPLEX64 = 8,
}


table Tensor {
  // 張量類型。每個項目的含義是特定於操作員的,但內建操作使用批次大小、高度、寬度、
通道數 (TensorFlow 的 NHWC)
  shape:[int];
  type:TensorType;
  // An index that refers to the buffers table at the root of the model. Or,
  // if there is no data buffer associated (i.e. intermediate results), then
  // this is 0 (which refers to an always existent empty buffer).
  //
  // The data_buffer itself is an opaque container, with the assumption that
the
```

```
  // target device is little-endian. In addition, all builtin operators assume
  // 記憶體的順序是這樣的：如果 shape 是 [4，3，2]，那麼索引 [i, j, k] 對映到資料緩衝
     區 [i*3*2 + j*2 + k]
  buffer:uint;
  name:string;   // 用於偵錯和匯入 TensorFlow
  quantization:QuantizationParameters;  // 可選
  is_variable:bool = false;
}


// 內建運算子列表。內建運算子比自訂運算子稍快
// 雖然自訂運算子接受包含設定參數的不透明物件，但內建函數有一組預先確定的可接受選項
enum BuiltinOperator : byte {
}


// 內建運算子的選項
union BuiltinOptions {
}


table Operator {
  // 索引到 operator_codes 陣列。在這裡使用整數可以避免複雜的地圖尋找
  opcode_index:uint;

  // 可選輸入和輸出張量用 -1 表示
  inputs:[int];
  outputs:[int];


  builtin_options:BuiltinOptions;
  custom_options:[ubyte];
  custom_options_format:CustomOptionsFormat;


  mutating_variable_inputs:[bool];
}
```

```
// 根類型，定義子圖，通常表示整個模型
table SubGraph {
  // 子圖中使用的所有張量的列表
  tensors:[Tensor];

  // 輸入此子圖的張量的索引。注意這是子圖進行推理的非靜態張量清單
  inputs:[int];

  // 此子圖輸出的張量的索引。注意這是子圖的推理
  outputs:[int];

  operators:[Operator];

  // 子圖的名稱（用於偵錯）
  name:string;
}

// 原始資料緩衝區表（用於常數張量）
table Buffer {
  data:[ubyte] (force_align: 16);
}

table Model {
  // 架構版本
  version:uint;

  // 此模型中使用的所有操作員程式的清單，它需要保持有序，因為運算子在其中攜帶索引向量
  operator_codes:[OperatorCode];

  // 模型的所有子圖。假設第 0 個是主模型
  subgraphs:[SubGraph];

  // 模型的描述 description:string;
```

```
// 模型緩衝區
// 注意，此陣列的第 0 個項目必須是空緩衝區（sentinel），這是一個約定，因此沒有緩衝
   區的張量可以提供 0 作為其緩衝區
buffers:[Buffer];

// 有關模型的中繼資料。間接進入現有緩衝區列表 metadata_buffer:[int];
}


root_type Model;
```

關於 FlatBuffers 的實際用法，我們來看看 TensorFlow Lite 是怎麼用的。在 TensorFlow 裡，可以參考 ensorflow/lite/model.h 和 ensorflow/lite/model.cc，這兩個檔案介紹了很多 FlatBuffers 的使用方法。

在 TensorFlow 裡主要使用了 VerifyAndBuildFromFile()，Android 的應用把模型的儲存路徑透過 JNI 傳進來，這個函數先讀取模型檔案，然後傳回一個 FlatBuffers 的模型，以下是相關的程式。

```
JNIEXPORT jlong JNICALL
Java_org_tensorflow_lite_NativeInterpreterWrapper_createModel(
    JNIEnv* env, jclass clazz, jstring model_file, jlong error_handle) {
  BufferErrorReporter* error_reporter =
      convertLongToErrorReporter(env, error_handle);
  if (error_reporter == nullptr) return 0;
  const char* path = env->GetStringUTFChars(model_file, nullptr);

  std::unique_ptr<tflite::TfLiteVerifier> verifier;
  verifier.reset(new JNIFlatBufferVerifier());

  auto model = tflite::FlatBufferModel::VerifyAndBuildFromFile(
      path, verifier.get(), error_reporter);
```

另外一個是 BuildFromBuffer()，與 VerifyAndBuildFromFile() 不同的是：
模型儲存在記憶體的 Buffer 裡，TensorFlow Lite 沒有直接使用這個函
數。但是如果開發者的應用已經把模型讀進了記憶體，就可以直接呼叫
這個函數。

在這裡我們可以特別關注一下 GetAllocationFromFile 函數，有兩個理
由：第一，這個實現是對應非 MCU 的，或説是為 Android 訂製的。第
二，如果 Android 支援 NNAPI，則選擇 NNAPI；如果核心支援共用記憶
體 MMAP，則選擇使用 MMAP。

```cpp
#ifndef TFLITE_MCU
// 從 "檔案名稱" 載入模型。如果 mmap_file 為 rue，則使用 mmap；不然在緩衝區中複製模型
std::unique_ptr<Allocation> GetAllocationFromFile(const char* filename,
                                                  bool mmap_file,
                                                  ErrorReporter*
                                                  error_reporter,
                                                  bool use_nnapi) {
  std::unique_ptr<Allocation> allocation;
  if (mmap_file && MMAPAllocation::IsSupported()) {
    if (use_nnapi && NNAPIDelegate::IsSupported())
      allocation.reset(new NNAPIAllocation(filename, error_reporter));
    else
      allocation.reset(new MMAPAllocation(filename, error_reporter));
  } else {
    allocation.reset(new FileCopyAllocation(filename, error_reporter));
  }
  return allocation;
}
```

當使用 MMAP 的時候，會呼叫 mmap()，這是 Posix 的 API，這個 API
的實現來自 Asylo。我們提起 Asylo 的原因是，機器學習的模型都非常
大，在行動裝置和嵌入式裝置上執行的時候，讀、寫和資料傳輸都會產

生效能上的問題，透過共用記憶體的方法來解決讀寫記憶體的問題是常用的解決方法，但是隨之會帶來安全上的問題，幸好 Asylo 為我們提供了一個解決方案，並作為 Posix 架構中的一部分。

Asylo 的官方定義是：

Asylo provides strong encapsulation around data and logic for developing and using an enclave. In the Asylo C++ API, an enclave application has trusted and untrusted components. The API has a central manager object for all hosted enclave applications.

Asylo 為開發和使用飛地提供了強大的資料和邏輯封裝。在 Asylo C++ API 中，安全區應用程式具有受信任和不受信任的元件。API 具有適用於所有託管安全區應用程式的中央管理員物件。

無論是英特爾的 SGX 還是 ARM TrustZone 都提供了一種隔離機制，用來保護程式和資料免遭修改或洩露。雖然我們使用的很多模型都是開放原始碼的，但是越來越多的模型將被開發出來，如果保護模型不被惡意侵犯，這其中既包含保護模型本身，也包含保護模型的執行環境，安全性會變得越來越重要。在國內，機器學習應用的差異化越來越高，應用的範圍越來越廣，對這方面的重視會與日俱增。

回到 FlatBuffers 本身，在 FlatBuffers 的 build_defs.bzl 裡，定義怎樣產生 FlatBuffers 的檔案，TensorFlow Lite 直接使用了它：

```
# Generic schema for inference on device.
flatbuffer_cc_library(
    name = "schema_fbs",
    srcs = ["schema.fbs"],
)
```

當編譯 TensorFlow Lite 的時候，schema 會被自動編譯，由 FlatBuffers 的編譯器在 bazel-out 的資料夾裡產生 C 程式和標頭檔，例如：

```
bazel-out/android-arm64-v8a-opt/genfiles/tensorflow/lite/schema/schema_
generated.h
```

這個檔案可以直接被參考，請注意它的字首路徑。

```
#include "tensorflow/lite/schema/schema_generated.h"
```

在這個標頭檔案中，GetModel 可以根據輸入的快取資料產生一個 FlatBuffers 的 Model，這個函數在 Lite 中被直接使用。關於 FlatBuffers 在 Lite 裡的實際應用，我們就不贅述了，大家可以參照原始程式碼。

```
inline const tflite::Model *GetModel(const void *buf) {
  return flatbuffers::GetRoot<tflite::Model>(buf);
}
```

FlatBuffers 的一些優缺點：

FlatBuffers 沒有解析步驟，這表示解析是隨選完成的。對於字串，這可能非常昂貴且緩慢。這在 Android 和行動裝置上是非常明顯的問題，其中在 UI 執行緒上的慢操作可能導致遺失畫面。

FlatBuffers 不提供資料封裝，因此不能儲存的空間將垃圾回收部分地作為後備緩衝區，這表示如果 FlatBuffers 中的任何衍生物件持續駐留，則整個後備資料陣列也是如此。

FlatBuffers 的核心 schema 會隨著程式無限增長，如果將來想要刪除一個欄位，那是不可能的。

FlatBuffers 的修改更昂貴，需要複製整個後備陣列。

FlatBuffers 的解析和序列化的速度要快得多。

大概歸納一下就是，FlatBuffers 產生的程式和函數庫程式都比 Protobuf 小一個數量級或更小，由 FlatBuffers 內建的線上或離線 JSON 解析和產生，是 FlatBuffers 與 gRPC 的無縫整合。

但這並不表示 FlatBuffers 比 Protobuf 更差，反之亦然。由於設計目標不同導致不同的權衡而設計出不同的資料結構。讀者需要根據自己的應用採用更好的資料結構，並且在同一應用程式中，以及在不同的執行環境中使用最合適的資料結構。

6.1.3 模型結構

一般來說我們需要先在桌上型電腦上設計、訓練出目標模型，並將其轉化成 TensorFlow Lite 的格式。接著，此格式檔案在 TensorFlow Lite 中會被內建 Neon 指令集的解析器載入到記憶體，並執行對應的計算。由於 TensorFlow Lite 對硬體加速介面有良好的支援，讀者可以設計出效能更優的 App 供使用者使用。在這裡，我們看看 TensorFlow Lite 裡的模型檔案格式。TensorFlow Lite 定義了 Model 這樣的結構，它是模型的主結構，實際程式如下：

```
table Model {
   version: uint;
   operator_codes: [OperatorCode];
   subgraphs: [SubGraph];

   description: string;
   buffers: [Buffer]
}
```

在上面的 Model 結構定義中，operator_codes 定義了整個模型的所有運算元，subgraphs 定義了所有的子圖。在子圖當中，第一個元素是主圖。

buffers 則是資料儲存區域，主要儲存模型的加權資訊。Model 中最重要的部分是 SubGraph，它也是一個結構：

```
table SubGraph {
    tensors: [Tensor];
    inputs: [int];
    outputs: [int];
    operators: [Operator];

    name: string;
}
```

在 SubGraph 裡定義了 Tensor，這也是一個結構，包含維度、資料類型、Buffer 位置等資訊。同理，ensors 定義了子圖的各個 Tensor，而 inputs 和 outputs 用索引的方法維護著子圖中 Tensor 與輸入輸出之間的對應關係。operators 定義了子圖當中的運算元。

```
table Tensor {
    shape: [int];
    type: TensorType;
    buffer: uint;

    name: string;
}
```

Buffer 以索引量的形式，列出這個 Tensor 需要用到子圖的哪個 Buffer。

在 SubGraph 中另一個重要的結構是 Operator，Operator 定義了子圖的結構：

```
table Operator {
    opcode_index: uint;
    inputs: [int];
    outputs: [int];
}
```

opcode_index 用索引的方式指明該 Operator 對應了哪個運算元。inputs 和 outputs 則是 Tensor 的索引值，指明該 Operator 的輸入輸出資訊。

6.1.4 轉換器（Toco）

由於 TensorFlow 使用了新的檔案格式和儲存模式，TensorFlow Lite 也提供了工具（Toco），讀者可以使用它進行模型的轉換。這是一個非常重要的工具，它負責把 TensorFlow 的模型轉換成 TensorFlow Lite 的模型。下面，讓我們來看一下 Toco 是運行原理的。

Toco 的程式位於 ensorflow/lite/toco 資料夾下。Toco 有三個主要功能，即匯入、匯出和轉換。匯入將輸入轉為 Model 類別。匯出將模型轉為 flite 模型或 graphviz。轉換以輸入標示為基礎對模型操作，並且它會刪除未使用的運算元等。下面是 Toco 的入口函數。

```
void ToolMain(const ParsedTocoFlags& parsed_toco_flags,
            const ParsedModelFlags& parsed_model_flags) {
            ModelFlags model_flags;
            ReadModelFlagsFromCommandLineFlags(parsed_model_flags,
&model_flags);

  TocoFlags toco_flags;
  ReadTocoFlagsFromCommandLineFlags(parsed_toco_flags, &toco_flags);

  string graph_def_contents;
  ReadInputData(parsed_toco_flags, parsed_model_flags, &toco_flags,
            &model_flags, &graph_def_contents);
            CheckOutputFilePermissions(parsed_toco_flags.output_file);

  std::unique_ptr<Model> model =
    Import(toco_flags, model_flags, graph_def_contents);
  Transform(toco_flags, model.get());
```

```
string output_file_contents;
Export(toco_flags, *model, toco_flags.allow_custom_ops(),
    &output_file_contents);
CHECK(port::file::SetContents(parsed_toco_flags.output_file.value(),
                              output_file_contents, port::file::Defaults())
                    .ok());
}
```

Toco 要做的事其實很簡單，就是把模型讀取然後輸出需要的模型。不過，如果我們了解了這個過程，就會對 TensorFlow Lite 的工作過程有更深入的了解。下面我們來看一下實作方式。

1 模型讀取

它基本上讀取 TensorFlow GraphDef 或 TensorFlow Lite 模型，然後轉為張量流模型。分析程式時，一個很有用的工具是 Kythe，使用它可以很快了解程式，也可以快速跳躍，筆者一般會把 TensorFlow 重新編譯，便於瀏覽程式，如圖 6-2 所示。Kythe 比一般 IDE 內建的程式索引的準確率要高，推薦給有興趣的讀者嘗試。

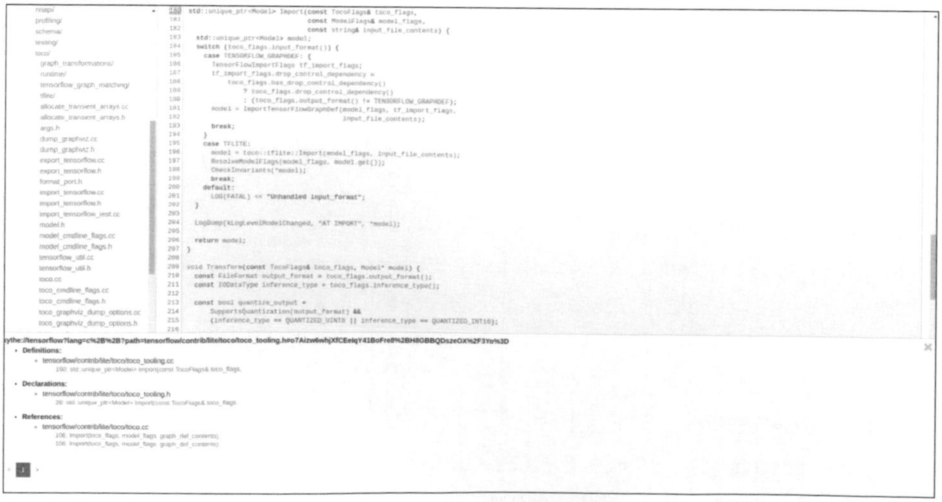

圖 6-2 Kythe 程式解析圖

例如下面的程式：

```
const ::tflite::Model* input_model =
     ::tflite::GetModel(input_file_contents.data());
```

你可以點擊 Model，就會跳躍到對應的 FlatBuffers 產生的程式中，這在很多 IDE 裡是做不到的，產生的程式如圖 6-3 所示。

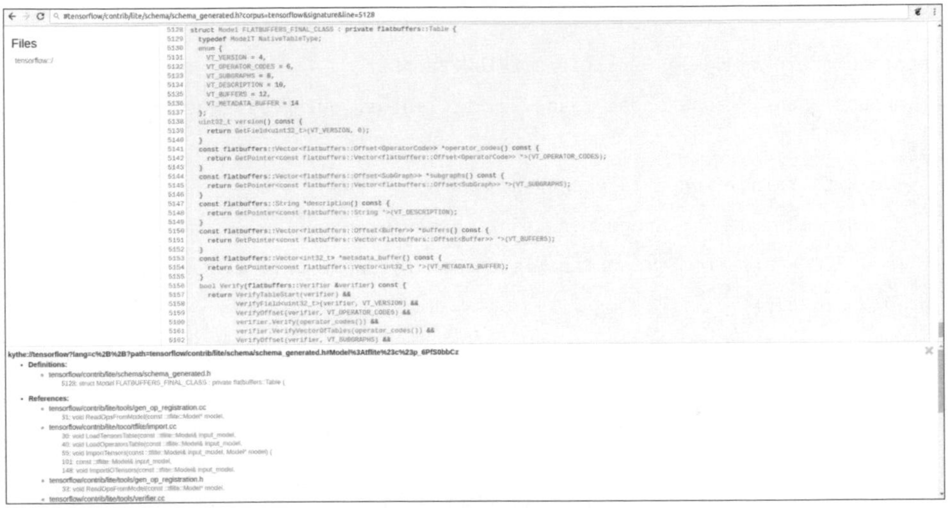

圖 6-3 產生程式圖

請不要和另一個 model.cc 混淆，它定義了一些輔助函數來複製模型。

現在我們回到 Toco。Toco 主要由兩個類別來實現。一個類別是對應 TensorFlow Lite 的，它是 FlatBuffers 的表示，另一個類別的程式在 Toco 資料夾中，它實現了 Toco 的內部邏輯。讓我們來說明 Toco 如何將張量流 graphdef 轉為 TensorFlow lite 模型，Toco 主要依靠下面幾個函數讀取原始模型檔案：

```
ImportTensors(*input_model, model.get());
ImportOperators(*input_model, ops_by_name, tensors_table, operators_table,
               model.get());
```

```
ImportIOTensors(*input_model, tensors_table, model.get());
ImportTensorFlowGraphDef(
    const ModelFlags& model_flags, const TensorFlowImportFlags& tf_import_
flags,
    const string& input_file_contents);
```

我們首先來看一下主要函數 ImportTensorFlowGraphDef，它會讀取一個
GraphDef 檔案資料：

```
std::unique_ptr<Model> ImportTensorFlowGraphDef(
    const ModelFlags& model_flags, const TensorFlowImportFlags& tf_import_
flags,
    const string& input_file_contents) {
    std::unique_ptr<GraphDef> tf_graph(new GraphDef);
    CHECK(ParseFromStringEitherTextOrBinary(input_file_contents,
tf_graph.get()));

    std::unique_ptr<GraphDef> pruned_graph =
        MaybeReplaceCompositeSubgraph(*tf_graph);
    if (pruned_graph) {
      tf_graph = std::move(pruned_graph);
    }
    return ImportTensorFlowGraphDef(model_flags, tf_import_flags,
*tf_graph);
}
```

下面兩行程式可以把輸入資料轉化成 GraphDef Protobuf：

```
std::unique_ptr<GraphDef> tf_graph(new GraphDef);
CHECK(ParseFromStringEitherTextOrBinary(input_file_contents, tf_graph.get()));
```

ParseFromStringEitherTextOrBinary 在 oco_port.h 中的定義程式如下，它
必須解決 Google 內部建置系統和開放原始碼 Protobuf 依賴問題。

```
std::unique_ptr<Model> ImportTensorFlowGraphDef(
    const ModelFlags& model_flags, const TensorFlowImportFlags& tf_import_
flags,
        const GraphDef& tf_graph);
```

讀取一個 GraphDef 後，我們就要進行模式轉換了。下面的函數會先建立一個空模型，然後檢查 GraphDef 中的節點。對於每個節點，呼叫 ImportTensorFlowNode 將張量流節點轉為 model.h 中的 TensorFlow Lite 的運算元。

```
Model* model = new Model;

for (auto node : inlined_graph.node()) {
  StripZeroOutputIndexFromInputs(&node);
  auto status = internal::ImportTensorFlowNode(node, tf_import_flags, model);
  CHECK(status.ok()) << status.error_message();
}

ResolveModelFlags(model_flags, model);

StripCaretFromArrayNames(model);
AddExtraOutputs(model);
FixNoMissingArray(model);
FixNoOrphanedArray(model);
FixOperatorOrdering(model);
CheckInvariants(*model);
```

2 模型輸出

上一節我們介紹了如何將模型轉為 TensorFlow Lite 的模式。export_tensorflow.cc 中的 ExportTensorFlowGraphDef 函數可以輸出模型的資訊和結構，其定義如下：

```
void ExportTensorFlowGraphDef(const Model& model,
                              string* output_file_contents) {
  CHECK(output_file_contents->empty());
  GraphDef tensorflow_graph;
  ExportTensorFlowGraphDefImplementation(model, &tensorflow_graph);
  LogDumpGraphDef(kLogLevelModelChanged, "AT EXPORT", tensorflow_graph);
  CHECK(tensorflow_graph.SerializeToString(output_file_contents));
```

有興趣的讀者可以從這個函數開始，把實現的內容讀清楚。上面我們已經說明了如何讀取 FlatBuffers 檔案，TensorFlow Lite 的模型是儲存在 FlatBuffers 中的，這裡不再贅述。

3 凍結模型

這裡我們說明一下凍結模型（freeze_graph）的主要功能和它的工作原理。訓練後的 TensorFlow 的模型不能被 TensorFlow Lite 直接使用，一定要把它固定（Freeze）。固定的實質是把模型的參數也一同寫進同一個模型檔案中。它的核心部分是 convert_variables_to_ constants，在 python/framework/graph_util_impl.py 中定義如下：

```
def convert_variables_to_constants(sess,
                                   input_graph_def,
                                   output_node_names,
                                   variable_names_whitelist=None,
                                   variable_names_blacklist=None):
```

本質上，該函數將圖形和輸出節點名稱作為輸入參數，從節點追蹤找出子圖形，然後在該子圖形中用常數取代所有變數。如圖 6-4 所示解釋了 freeze_graph 中的基本邏輯。

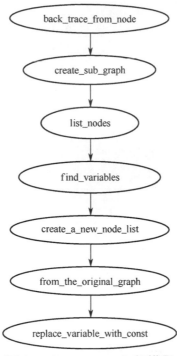

圖 6-4　freeze_graph 架構圖

下面的程式說明了實作方式的細節：

```
for input_node in inference_graph.node:
    output_node = node_def_pb2.NodeDef()
    if input_node.name in found_variables:
        output_node.op = "Const"
        output_node.name = input_node.name
        dtype = input_node.attr["dtype"]
        data = found_variables[input_node.name]
        output_node.attr["dtype"].CopyFrom(dtype)
        output_node.attr["value"].CopyFrom(
            attr_value_pb2.AttrValue(
                tensor=tensor_util.make_tensor_proto(
                    data, dtype=dtype.type, shape=data.shape)))
```

```
   how_many_converted += 1
elif input_node.op == "ReadVariableOp" and (
     input_node.input[0] in found_variables):
   # The preceding branch converts all VarHandleOps of ResourceVariables to
   # constants, so we need to convert the associated ReadVariableOps to
   # Identity ops.
   output_node.op = "Identity"
   output_node.name = input_node.name
   output_node.input.extend([input_node.input[0]])
   output_node.attr["T"].CopyFrom(input_node.attr["dtype"])
   if "_class" in input_node.attr:
     output_node.attr["_class"].CopyFrom(input_node.attr["_class"])
else:
   output_node.CopyFrom(input_node)
output_graph_def.node.extend([output_node])
```

6.1.5 解析器（Interpreter）

那麼 TensorFlow Lite 的解析器又是如何運行的呢？我們來學習一下。

一開始，終端裝置會透過 mmap 以記憶體對映的形式將模型檔案載入用戶端記憶體中，其中包含了所有的 Tensor、Operator 和 Buffer 等資訊。出於資料使用的需要，TensorFlow Lite 會同時建立 Buffer 的唯讀區域和分配寫入 Buffer 區域。由於解析器中包含了全部執行計算的程式，這一部分被稱為 Kernel。模型中的各個 Tensor 會被載入為 TfLiteTensor 的格式聯集中儲存在 TfLiteContext 中。每個 Tensor 的指標都指向記憶體中的唯讀 Buffer 區域，或是一開始新分配的寫入 Buffer 區域。模型中的 Operator 被重新載入為 TfLiteNode，它包含輸入輸出的 Tensor 索引值。這些 Node 對應的運算符號儲存於 TfLiteRegistration 中，它包含了指向 Kernel 的函數指標。OpResolver 負責維護函數指標的對應關係。

TensorFlow Lite 在載入模型的過程中會確定執行 Node 的順序，然後依次執行。

讀者如果想要更進一步地掌握 TensorFlow Lite 的技術細節，一定要閱讀以下檔案：

```
lite/context.h
lite/model.h
lite/interpreter.h
lite/kernels/register.h
```

程式執行的順序如表 6-1 所示。

表 6-1

程式	檔案路徑
run	tensorflow/java/src/main/java/org/tensorflow/Session.java
run	tensorflow/java/src/main/native/session_jni.cc
Session::Run	tensorflow/core/common_runtime/session.cc
run	tensorflow/core/common_runtime/direct_session.cc

DirectSession 是 Session 的子類別，定義如下：

```
class DirectSession : public Session;
```

解譯器是一個靜態變數，該變數在載入 TensorFlow 函數庫時被初始化，在其建置函數中，Session 是透過 SessionFactory::Register 註冊的。

```
class DirectSessionRegistrar {
 public:
  DirectSessionRegistrar() {
    SessionFactory::Register("DIRECT_SESSION", new DirectSessionFactory());
  }
};
static DirectSessionRegistrar registrar;
```

解析器是由原生程式（C++）產生的，下面我們來看一下解譯器的內部實現程式：

```cpp
JNIEXPORT jlong JNICALL
Java_org_tensorflow_lite_NativeInterpreterWrapper_createInterpreter(
    JNIEnv* env, jclass clazz, jlong model_handle, jlong error_handle,
    jint num_threads) {
  tflite::FlatBufferModel* model = convertLongToModel(env, model_handle);
  if (model == nullptr) return 0;
  BufferErrorReporter* error_reporter =
      convertLongToErrorReporter(env, error_handle);
  if (error_reporter == nullptr) return 0;
  auto resolver = ::tflite::CreateOpResolver();
  std::unique_ptr<tflite::Interpreter> interpreter;
  TfLiteStatus status = tflite::InterpreterBuilder(*model, *(resolver.
get()))(
      &interpreter, static_cast<int>(num_threads));
  if (status != kTfLiteOk) {
    throwException(env, kIllegalArgumentException,
                   "Internal error: Cannot create interpreter: %s",
                   error_reporter->CachedErrorMessage());
    return 0;
  }
  // 分配記憶體
  status = interpreter->AllocateTensors();
  if (status != kTfLiteOk) {
    throwException(env, kNullPointerException,
                   "Internal error: Cannot allocate memory for the
interpreter",
                   error_reporter->CachedErrorMessage());
    return 0;
  }
  return reinterpret_cast<jlong>(interpreter.release());
}
```

在解譯器實例的產生過程中包含了所有運算元實例的產生，而運算元的
產生是由多載組成的，對應的程式如下：

```
TfLiteStatus InterpreterBuilder::operator()(
    std::unique_ptr<Interpreter>* interpreter, int num_threads) {
  if (!interpreter) {
    error_reporter_->Report(
        "Null output pointer passed to InterpreterBuilder.");
    return kTfLiteError;
  }


  // 透過刪除部分解譯器安全退出，以減少冗長的內容
auto cleanup_and_error = [&interpreter]() {
    interpreter->reset();
    return kTfLiteError;
  };

  if (!model_) {
    error_reporter_->Report("Null pointer passed in as model.");
    return cleanup_and_error();
  }

  if (model_->version() != TFLITE_SCHEMA_VERSION) {
    error_reporter_->Report(
        "Model provided is schema version %d not equal "
        "to supported version %d.\n",
        model_->version(), TFLITE_SCHEMA_VERSION);
    return cleanup_and_error();
  }

  if (BuildLocalIndexToRegistrationMapping() != kTfLiteOk) {
    error_reporter_->Report("Registration failed.\n");
    return cleanup_and_error();
  }
```

```cpp
// FlatBuffers 模型的模式定義獨立於圖形的操作碼清單。我們將這些對映到登錄檔，以便
// 減少自訂的字串尋找操作，我們只對每個自訂操作執行一次
auto* subgraphs = model_->subgraphs();
  auto* buffers = model_->buffers();
  if (subgraphs->size() != 1) {
    error_reporter_->Report("Only 1 subgraph is currently supported.\n");
    return cleanup_and_error();
  }
  const tflite::SubGraph* subgraph = (*subgraphs)[0];
  auto operators = subgraph->operators();
  auto tensors = subgraph->tensors();
  if (!operators || !tensors || !buffers) {
    error_reporter_->Report(
        "Did not get operators, tensors, or buffers in input flat buffer.\n");
    return cleanup_and_error();
  }
  interpreter->reset(new Interpreter(error_reporter_));
  if ((**interpreter).AddTensors(tensors->Length()) != kTfLiteOk) {
    return cleanup_and_error();
  }
  // 設定 num 執行緒
(**interpreter).SetNumThreads(num_threads);
  // 分析輸入 / 輸出
  (**interpreter).SetInputs(FlatBufferIntArrayToVector(subgraph-> inputs()));
  (**interpreter).SetOutputs(FlatBufferIntArrayToVector(subgraph-> outputs()));

  // 最後設定節點和張量
if (ParseNodes(operators, interpreter->get()) != kTfLiteOk)
    return cleanup_and_error();
  if (ParseTensors(buffers, tensors, interpreter->get()) != kTfLiteOk)
    return cleanup_and_error();

  return kTfLiteOk;
}
```

解譯器產生之後即可啟用了：

```
if (interpreter->Invoke() != kTfLiteOk)
```

6.2 底層結構和設計

TensorFlow Lite 提供了 C++ 和 Java API，在這兩套 API 裡，API 設計都特別強調了便利性。TensorFlow Lite 是專為行動裝置和小型裝置上的快速推理而設計的，TensorFlow Lite 希望為開發人員提供簡單而高效的 C++ API，以便在 TensorFlow Lite 模型上高效執行推理計算。

6.2.1 設計目標

TensorFlow Lite 是一種推理引擎（Inference Engine），可用於在包含行動裝置在內的小型裝置上執行。開發者要將 SavedModel 或凍結後的 GraphDef 轉為 TensorFlow Lite 自己的 FlatBuffers 格式。

在使用 TensorFlow Lite 前，開發者需要一種簡單的方法來載入他們的模型，將資料提供給推理引擎並獲得結果。TensorFlow 的很多程式可以直接在行動裝置上執行，但是 TensorFlow Lite 簡化了開發者的工作流程，並且能和 TensorFlow 及其他開放原始碼架構直接進行協調和工作。

應該注意的是，TensorFlow Lite 通常針對小型裝置，而不僅是行動裝置。這裡很難定義小型裝置，可以基本認為 TensorFlow Lite 都可以適用於不在資料中心和作為桌面計算裝置以外的任何場景。TensorFlow Lite 本身的檔案佔用空間很小，執行佔用的空間也比較小。另外，TensorFlow Lite 對外部函數庫的依賴性也很小。

為了在 TensorFlow Lite 中執行推理模型，使用者需要將模型載入到
FlatBufferModel 物件中，然後由 Interpreter 執行。FlatBufferModel 需要
保持在 Interpreter 的整個生命週期內有效，並且單獨的 FlatBufferModel
可以由多個 Interpreter 同時使用。實際而言，必須在使用它的任何解譯
器物件之前建立 FlatBufferModel 物件，並且不釋放該物件，直到它們全
部被銷毀。

TensorFlow Lite 的最簡單的用法如下所示：

```
std::unique_ptr<tflite::FlatBufferModel> model = tflite::FlatBufferModel::
BuildFromFile(path_to_model);

tflite::ops::builtin::BuiltinOpResolver resolver;
tflite::Interpreter interpreter = tflite::Interpreter::Create(model, resolver);
interpreter->SetInput(0, std::vector<float>({...}));
interpreter->Invoke(); const Tensor* output = interpreter-> GetOutputTensor(0);
```

在上面的程式片段中，我們已經體會到一些 TensorFlow Lite 的設計概要：

- Ops 和 kernels 之間的連接由 OpResolver 提供，如果需要可以使用它對
  Ops 進行置換，允許開發者只包含他們所需要的核心。TensorFlow Lite
  有自己的核心，這個核心的依賴很少，因此開發人員可以使用自己的
  OpResolver 建立非常小的可執行檔。

- 輸入和輸出張量（Tensor）由整數表示，而非字串。在大多數情況
  下，這是為了避免處理字串操作，因此程式佔用空間很小。一般來説
  使用 TensorFlow Lite 的開發人員已經離線處理了他們的模型，並且確
  切地知道它們的輸入和輸出張量的順序。對於不關心字串處理程式的
  讀者，可以使用其名稱尋找到相對應的張量。

- 無法只執行圖形的一部分。出於效率原因，TensorFlow Lite 假設圖形
  是按執行順序定義的，並且所有節點對於推理都是絕對必要的。

■ 透過 Tensor 提供對輸出張量的存取。在程式內部，TensorFlow Lite 張量是純 C 結構的，因此核心實現者可以在不依賴 C++ 的情況下實現操作。TensorFlow Lite 簡化了對內部張量資料相關部分的存取。

6.2.2 錯誤回饋

為了保持較小的二進位大小，TensorFlow Lite 避免依賴於 std::stream 和進階字串函數庫。在許多地方，TensorFlow Lite 透過簡單的 TfLiteStatus 物件傳回狀態資訊，對應的程式如下：

```
typedef enum {
    kTfLiteOk = 0, kTfLiteError = 1 } TfLiteStatus;
    Failures can be easily verified:
    if (status != kTfLiteOk) {      // 此處處理錯誤
}
```

但進一步的顯示出錯需要更多的處理。在已部署的應用程式中，我們不希望觸發任何錯誤處理程式，但是使用者仍需要檢視錯誤訊息以進行偵錯，TensorFlow Lite 為詳細的錯誤報告提供了類似 printf 的介面，相關程式如下：

```
class ErrorReporter {
    virtual int Report(const char* format, va_list args) = 0;
};
```

為了向 stderr 報告錯誤，TensorFlow Lite 提供了 DefaultErrorReporter，開發者可以從 ErrorReporter 衍生出來。

6.2.3 載入模型

FlatBufferModel 類別封裝了一個模型，可以根據模型的儲存類型和位置，採用靈活的方式進行建置，實作方式程式如下：

```
class FlatBufferModel {
    // 基於檔案建置模型。如果失敗,傳回 nullptr
static std::unique_ptr<FlatBufferModel> BuildFromFile(
    const char* filename, ErrorReporter* error_reporter);
    // 以預先載入為基礎的 FlatBuffer 建置模型。呼叫方保留緩衝區的所有權,並應保持緩
衝區的活動狀態,直到傳回的物件被銷毀。如果失敗,傳回 nullptr
static std::unique_ptr<FlatBufferModel> BuildFromBuffer(
    const char* buffer, size_t buffer_size, ErrorReporter* error_reporter);
    // 以現有為基礎的檔案描述符號建置模型。如果失敗則傳回 nullptr
 static std::unique_ptr<FlatBufferModel> BuildFromFileDescriptor(
    int file_descriptor, size_t buffer_size, ErrorReporter* error_reporter);
};
```

當模型載入檔案時,如果 TensorFlow Lite 檢測到 Android NNAPI,它將
自動嘗試使用共用記憶體來儲存 FlatBufferModel。確保 NNAPI 可用的
使用者可以傳遞表示共用記憶體的檔案描述符號。Java 使用者很可能將
他們的模型載入到位元組緩衝區並將它們直接傳遞給 C++ API。

6.2.4 執行模型

執行模型需要幾個簡單的步驟:

- 以現有為基礎的 FlatBufferModel 建置解譯器。

- 選擇調整輸入張量的大小和形狀。出於效率原因,TensorFlow Lite 模
  型通常已經包含所有張量的大小和形狀。

- 設定輸入張量值,方法是將資料複製到解譯器中或參考外部分配的資
  料。

- 呼叫推理。

- 讀取輸出張量值。

- 可選擇重複最後三個步驟：設定輸入、呼叫、讀取輸出。

解譯器（Interpreter）的公共介面開發程式如下：

```
class Interpreter {

    static TfLiteStatus Create(const FlatBufferModel* model,
    const OpResolver& op_resolver, ErrorReporter* error_reporter);

    // 對輸入索引清單的唯讀存取
    const std::vector<size_t>& inputs() const;

    // 對輸出索引清單的唯讀存取
const std::vector<size_t>& outputs() const;

    // 改變指定張量的維數。" 張量索引 " 應該是 inputs() 傳回的索引之一
    TfLiteStatus ResizeInputTensor(size_t tensor_index, const
std::vector<int>& dims);

    // 將資料複製到輸入張量中
    template <typename T> TfLiteStatus SetInputTensor(size_t tensor_index,
std::vector<T> data);

    // 將輸入張量的值設定為對外部分配的參考記憶體

    template <typename T> TfLiteStatus SetInputTensorFromMemory(size_t
tensor_index, T* buffer, size_t buffer_size_in_bytes);

    // 表示外部分配的記憶體緩衝區的半不透明類型。這在嘗試使用 NNAPI 緩衝區設定張量資
料時特別有用
class MemoryBuffer {
```

```
    enum Type {
        ANDROID_SHARED_MEMORY = 0;
    };

    virtual Type GetType() = 0;
```

// 將資料複製到指定的緩衝區中。如果複製失敗，則傳回 false
```
virtual bool CopyTo(void* buffer, size_t buffer_size_in_bytes) = 0;
```

```
    // 傳回記憶體緩衝區中的位元組數
    virtual size_t Size() = 0;
    };
```

```
    // 設定輸入張量的值作為對記憶體緩衝區的參考。當後端執行 Ops 時，可以將緩衝區類型
強制轉換成內部記憶體類型，不進行複製
TfLiteStatus SetInputTensorFromBuffer(size_t tensor_index, MemoryBuffer*
buffer);
```

```
// 傳回一個指標，該指標指向儲存在輸出張量中的值
// 如果 ensor_index 超出範圍或不是輸出張量，則傳回空指標
// 對 getOutputtensor() 的呼叫和傳回的張量僅在 invoke() 之後有效
const Tensor* GetOutputTensor(size_t tensor_index);
```

```
// 傳回 memorybuffer
MemoryBuffer GetOutputTensor(size_t tensor_ index);
```

```
// 執行模型，填充輸出張量。將會使以前從 getOutputEnsor（）取得的參考無效
// 從 GetOutputTensor（）取得的參考
    TfLiteStatus Invoke();
    };
```

解譯器具有以下特性：

■ 張量由整數表示，以避免字串比較，以及對字串函數庫的任何固定依賴。

■ 可以設定輸入值而不進行複製。

■ 存取輸出資料不會曝露內部 TensorFlow Lite 的張量表示，而是傳回包裝器物件。

■ 不能從平行處理執行緒存取解譯器。

■ 在調整張量大小後立即分配內部和輸出張量的記憶體。

6.2.5 訂製演算子（CUSTOM Ops）

除了 TensorFlow Lite 附帶的演算子，讀者還可以訂製演算子。由於現在 TensorFlow Lite 附帶的演算子非常有限，如果模型變得越來越複雜，可能一段時間內，讀者要訂製很多演算子。

下面我們來看一下 TensorFlow Lite 是怎樣實現的。

TensorFlow Lite 的目標之一是為人們提供建置解譯器的基本部分，在這種開發環境裡，沒有完整的 C++ 工具鏈，如果讀者的建置目標是 DSP 或其他類型的裝置，只能使用 C。TensorFlow Lite 架構提供了幾個可用於撰寫自訂操作的 C 結構：

■ TfLiteContext 提供對解譯器狀態的存取，可用於檢索全域物件，包含張量。更常見的是，TfLiteContext 用於報告操作員處理中的錯誤。

■ TfLiteNode 包含有關正在執行的操作的資訊。實現可以使用此物件存取其輸入和輸出張量。

下面是 **TfLiteContext** 和 **TfLiteNode** 的定義，程式如下：

```
struct TfLiteContext {

    // 模型中張量的數目：int tensors_size;
    // 模型中張量的清單：TfLiteTensor* tensors;
    // 更新張量的維數
    TfLiteStatus (*ResizeTensor)(struct TfLiteContext*, TfLiteTensor* tensor,

    TfLiteIntArray* new_size);
    // 請求用格式字串 msg 報告錯誤
    void (*ReportError)(struct TfLiteContext*, const char* msg, ...);

    // 增加 ensors_to_add 張量，保留現有的張量。如果 ensors_to_add 的值為不可為空，
則 First_New_Tensor_Index 指向的值將設定為第一個新張量的索引
    TfLiteStatus (*AddTensors)(struct TfLiteContext*, size_t tensors_to_add,

    int* first_new_tensor_index); };

    struct TfLiteNode {

    // 此節點的輸入表示為 TfLiteContext 張量的索引
TfLiteIntArray* inputs;

    // 輸出到該節點，表示為 TfLiteContext 張量的索引
    TfLiteIntArray* outputs;

    // 該節點的 init() 函數傳回的不透明資料（見下文）
    void* user_data;

    // 在輸入 FlatBuffers 時為該節點提供的不透明資料。這只適用於內建操作
    void* builtin_data;
};
```

撰寫 TensorFlow Lite 核心有關定義四個 C 函數：init()、prepare()、invoke() 和 free()，無論是自訂還是內建操作都是同樣的，相關程式如下：

```
typedef struct _TfLiteRegistration {

// 初始化序列化資料中的操作
// 如果是內建 OP：buffer 操作的參數資料（flite1stmparams*）的長度為零
// 如果是自訂 OP：buffer 是操作的 " 自訂 " 選項
// length 是緩衝區的大小
// 傳回類型 punned（即 void*）是不透明資料（如基元指標或結構的實例）
// 傳回的指標將與節點一起儲存在 user_data 欄位中，可在下面的準備和呼叫函數中存取
// 注意：如果資料已經是所需的格式，只需實現傳回 nullptr 的函數，並將自由函數實現為
no-op
  void* (*init)(TfLiteContext* context, const char* buffer, size_t length);

  // 指標 buffer 是先前由 init 呼叫傳回的資料
  void (*free)(TfLiteContext* context, void* buffer);

  // 當此節點所依賴的輸入已調整大小時，將呼叫 Prepare
  // 可以透過呼叫 context->resizetensor() 來請求調整輸出張量的大小

  // 成功傳回 ktfliteok
  TfLiteStatus (*prepare)(TfLiteContext* context, TfLiteNode* node);

  // 執行節點 ( 應該取 node->inputs 和 output 到 node->outputs)
  // 成功傳回 kTfLiteOk
  TfLiteStatus (*invoke)(TfLiteContext* context, TfLiteNode* node);

  // 在概要分析資訊期間呼叫概要分析字串，以便使執行分組在一起
 const char* (*profiling_string)(const TfLiteContext* context,
                                 const TfLiteNode* node);
```

```
// 內建程式。注意：註冊綁定器負責設定要正確
  int32_t builtin_code;

  // Custom op name. If the op is a builtin, this will be null.
  // Note: It is the responsibility of the registration binder to set this
  // properly.
  // WARNING: This is an experimental interface that is subject to change.
  // 自訂操作名稱。如果 OP 是內建的，那麼它將為空
  // 注意：要正確設定註冊綁定器
  // 注意：這是一個可更改的實驗介面
  const char* custom_name;

  // 操作的版本
  int version;
} TfLiteRegistration;
```

當解譯器載入模型時，它會為圖中的每個節點呼叫一次 init()。這表示如果在圖中多次使用指定的 init() 將被多次呼叫。init() 的宣告方式如下：

```
void* (*init)(TfLiteContext* context, const char* buffer, size_t length);
```

使用 FlexBuffers 的注意事項：自訂操作的緩衝區不必是 FlexBuffer。我們將使用列舉來定義每個自訂緩衝區，該列舉描述它包含的資料類型：

```
enum CustomOpDataType {
    FLEXBUFFER = 0;
};
```

如果我們願意，可以在未來使用不同的類型資料作為緩衝。對於每個 init() 呼叫，都會有一個相對應的 free() 的呼叫，允許實現處理它們在呼叫 init() 時傳回的緩衝區：

```
void (*free)(TfLiteContext* context, void* init_data);
```

每當輸入張量調整大小時，解譯器將透過圖表通知變更。這使他們有機會調整內部緩衝區的大小，檢查輸入形狀和類型的有效性，並重新計算輸出形狀。處理此階段的函數程式如下：

```
TfLiteStatus (*prepare)(TfLiteContext* context, TfLiteNode* node);
```

另外，每次推理執行時期，解譯器會檢查圖形且呼叫 invoke()：

```
TfLiteStatus (*invoke)(TfLiteContext* context, TfLiteNode* node);
```

custom 和 builtin Ops 都提供了全域註冊功能，定義如下所示：

```
namespace tflite {
    namespace ops {
        namespace builtin {
            TfLiteRegistration* Register_MY_CUSTOM_OP() {
                static TfLiteRegistration r = { my_custom_op::Init,
my_custom_op::Free, my_custom_op::Prepare, my_custom_op::Eval};
                return &r;
            }
        } // namespace builtin
    } // namespace Ops
} // namespace tflite
```

請注意，註冊不是自動的，應該在某處顯性呼叫 Register_MY_CUSTOM_OP。

例如下面的程式要註冊一個叫 MY_CUSTOM_OP 的運算元：

```
namespace tflite {
namespace ops {
namespace custom {
  TfLiteRegistration* Register_MY_CUSTOM_OP() {
    static TfLiteRegistration r = {my_custom_op::Init,
                                   my_custom_op::Free,
```

```
                                my_custom_op::Prepare,
                                my_custom_op::Eval};

    return &r;
  }
}  // namespace custom
}  // namespace Ops
}  // namespace tflite
```

然後，呼叫註冊函數。TensorFlow Lite 是在 BuiltinOpResolver::Builtin
OpResolver() 中註冊運算子的，例如下面的程式註冊了五個訂製的運算
元：

```
AddCustom("Mfcc", tflite::ops::custom::Register_MFCC());
AddCustom("AudioSpectrogram",
        tflite::ops::custom::Register_AUDIO_SPECTROGRAM());
AddCustom("LayerNormLstm", tflite::ops::custom::Register_LAYER_NORM_LSTM());
AddCustom("Relu1", tflite::ops::custom::Register_RELU_1());
AddCustom("TFLite_Detection_PostProcess",
        tflite::ops::custom::Register_DETECTION_POSTPROCESS());
```

這樣我們就完成了註冊訂製運算元的過程。

6.2.6 訂製核心

實作方式時，解譯器將載入一個核心函數庫，這些核心將被分配用於執
行模型中的每個運算符號。雖然預設函數庫僅包含內建核心，但可以使
用自訂函數庫取代它。

解譯器使用 OpResolver 將操作程式和名稱轉為實際程式：

```
class OpResolver {

    virtual TfLiteRegistration* FindOp(tflite::BuiltinOperator op) const = 0;
```

```
virtual TfLiteRegistration* FindOp(const char* op) const = 0;
virtual void AddOp(tflite::BuiltinOperator op,

TfLiteRegistration* registration) = 0;
virtual void AddOp(const char* op, TfLiteRegistration* registration) = 0;
};
```

最常見的用法僅依賴於 BuiltinOperator 列舉定義操作名稱，避免使用
'const char *。在某些應用程式中，避免字串比較和 std::set 很重要。自訂
核心必然會產生字串尋找，如果這些核心的數量很少，尋找速度也會很
快。在任何情況下，在解譯器初始化時，每個節點都會發生一次尋找。

如果開發人員希望在現有的解析器中增加一個或兩個自訂操作。他們可
以使用 BuiltinOpResolver 並手動註冊自訂操作，如下所示：

```
tflite::ops::builtin::BuiltinOpResolver resolver;
TfLiteStatus status = resolver.AddOp("my_custom_op", Register_MY_CUSTOM_
OP());
```

註冊已存在的操作是錯誤的。如果內建操作集被認為太大，則可以以指
定為基礎的操作子集用程式產生新的 OpResolver。

6.3 工具

由於 TensorFlow Lite 和其他 TensorFlow 元件比較起來比較新，有幾個
工具非常有用，可以幫助讀者了解 TensorFlow Lite，下面我們介紹兩個
有用的工具。

6.3.1 影像標記（label_image）

這個工具在 ensorflow/lite/examples/label_image 裡，它是一個用 C++ 寫的影像分類工具。它會讀取一個影像和模型檔案，按照標記的檔案進行分類。

首先，下載影像標記要用到的幾個檔案，程式如下：

```
$ ./tensorflow/lite/examples/ios/download_models.sh
download_models.sh 檔案的功能是下載以下幾個檔案：
./tensorflow/lite/examples/ios/simple/data/mobilenet_v1_1.0_224.tflite
./tensorflow/lite/examples/ios/camera/data/mobilenet_quant_v1_224.tflite
./tensorflow/lite/examples/ios/camera/data/mobilenet_v1_1.0_224.tflite
```

然後，建置執行檔案。Android 裝置的 ABI 可以透過以下指令獲得：

```
adb shell getprop | grep abi
```

假設你的裝置是 armv864 位元晶片，可以執行下面的指令：

```
$ bazel build --config android_arm64 --config monolithic --cxxopt=-std=c++11 \
  //tensorflow/lite/examples/label_image:label_image
```

注意，我們需要 --config monolithic，不然會出現編譯問題。

編譯後會獲得 bazel-bin/tensorflow/lite/examples/label_image/label_image，然後我們把它和必要的檔案直接拷貝到一台 Android 手機上，並執行下面的指令：

```
# 複製執行檔案
$ adb push bazel-bin/tensorflow/lite/examples/label_image/label_image/data/
local/tmp
bazel-bin/tensorflow/lite/examples/label_image/label_image: 1 file pushed.
19.5 MB/s (1751824 bytes in 0.086s)
```

```
# 複製模型檔案
$ adb push ./tensorflow/lite/examples/ios/camera/data/mobilenet_quant_
v1_224.tflite /data/local/tmp
./tensorflow/lite/examples/ios/camera/data/mobilenet_quant_v1_224.tflite: 1
file pushed. 18.9 MB/s (4276100 bytes in 0.216s)

# 複製測試影像檔
$ adb push ./tensorflow/lite/examples/label_image/testdata/grace_ hopper.bmp
/data/local/tmp
./tensorflow/lite/examples/label_image/testdata/grace_hopper.bmp: 1 file
pushed. 13.8 MB/s (940650 bytes in 0.065s)

# 複製標記結果檔案
$ adb push ./tensorflow/lite/examples/ios/simple/data/labels.txt /data/
local/tmp
./tensorflow/lite/examples/ios/simple/data/labels.txt: 1 file pushed. 0.7
MB/s (10484 bytes in 0.014s)
```

下面，在 Android 的手機上執行指令 "$adb shell"，把執行環境從桌面跳躍到手機，此時可以看到提示符號 "# taimen"，這是筆者的手機的名稱。手機的品牌不同，提示符號也會不同。

如果讀者是 Android 的開發者，對解鎖（Unlock）應該不陌生。由於我們要向手機裝置裡拷貝檔案，需要額外的許可權，所以手機需要先解鎖。由於手機和電信業者不同，解鎖的過程不同，讀者需要自己去了解解鎖過程。

現在，透過執行指令 "aimen:/ # cd /data/local/tmp" 把目前路徑傳輸到 /data/local/tmp 下，因為在這個路徑下，我們可以執行 Android 的應用。

執行指令 "aimen:/ # ls -all"，獲得以下檔案列表：

```
drwxrwx--x 3 shell shell    40962019-01-1501:35:11.651012625 -0500 .
drwxr-x--x 4 root  root     40962019-01-0718:26:05.842333353 -0500 ..
drwxrwxrwx 5 shell shell    40962019-01-0719:34:27.222741927 -0500 deployment
-rw-rw-rw- 1 root  root   9406502019-01-1414:24:27.000000000 -0500 grace_
hopper.bmp
-r-xr-xr-x 1 root  root  17518242019-01-1500:56:19.000000000 -0500 label_
image
-rw-rw-rw- 1 root  root    104842019-01-1501:05:16.000000000 -0500 labels.txt
-rw-rw-rw- 1 root  root  42761002019-01-1501:05:16.000000000 -0500
mobilenet_quant_v1_224.tflite
```

然後，執行 label_image 指令稿來檢測影像識別的準確度，執行結果如下：

```
/aimen:/data/local/tmp # ./label_image
WARNING: linker: Warning: "/data/local/tmp/label_image" unused DT entry:
DT_RPATH (type 0xf arg 0x488) (ignoring)
Loaded model ./mobilenet_quant_v1_224.tflite
resolved reporter
invoked
average time: 79.017 ms
0.666667: 458 bow tie
0.290196: 653 military uniform
0.0117647: 835 suit
0.00784314: 611 jersey
0.00392157: 922 book jacket
```

這些工具非常有用。因為 Android 開發基本都用 Java，所以我們要寫 Java 的應用並使其與 JNI 連接起來。可是，TensorFlow 都是透過 C++ 實現的，在對效能要求高的裝置上，用 C++ 的工具去測試會方便很多，加強工作效率。舉例來説，可以先用這個工具去測試效能，可以省去很多不必要的 Android 操作，而且可以測試更多的場景。我們看到了，這個工具是用 C++ 寫的，那麼它經過編譯以後，應該可以在各種平台上執行。在以 *nix 為基礎的平台上，例如 Ubuntu、CentOS、Mac 上也可以執行。

下面來看一下這個工具的一些有用的功能。由於這些功能是共通的，我們也可以在本機機上執行這些功能。在 X86 的 Ubuntu、Mac 或類似的機器上執行以下程式：

```
$ bazel run --cxxopt=-std=c++11 //tensorflow/lite/examples/label_image -- -h
label_image
--accelerated, -a: [0|1], use Android NNAPI or not
--count, -c: loop interpreter->Invoke() for certain times
--input_mean, -b: input mean
--input_std, -s: input standard deviation
--image, -i: image_name.bmp
--labels, -l: labels for the model
--tflite_model, -m: model_name.tflite
--profiling, -p: [0|1], profiling or not
--num_results, -r: number of results to show
--threads, -t: number of threads
--verbose, -v: [0|1] print more information
```

由於這個工具實際沒有 -h 幫助選項，我們需要讓程式觸發錯誤，然後產生説明選單。我們可以用下面三個選項指定不同的輸入檔案：

```
--image, -i: image_name.bmp
--labels, -l: labels for the model
--tflite_model, -m: model_name.tflite
```

舉例來説，可以把模型，標記檔案和圖形檔案都儲存在本機：

```
$ bazel-bin/tensorflow/lite/examples/label_image/label_image -i ./grace_
hopper.bmp -l ./labels.txt -m ./mobilenet_quant_v1_224.tflite
```

如果現在使用詳細輸出選項（Verbose），那麼輸出結果應該是這樣的：

```
$ bazel-bin/tensorflow/lite/examples/label_image/label_image -i ./grace_
hopper.bmp -l ./labels.txt -m ./mobilenet_quant_v1_224.tflite -v 1
Loaded model ./mobilenet_quant_v1_224.tflite
resolved reporter
tensors size: 89
nodes size: 31
inputs: 1
input(0) name: Placeholder
0: MobilenetV1/Logits/AvgPool_1a/AvgPool, 1024, 3, 0.0235285, 0
1: MobilenetV1/Logits/Conv2d_1c_1x1/BiasAdd, 1001, 3, 0.165351, 74
2: MobilenetV1/Logits/Conv2d_1c_1x1/Conv2D_bias, 4004, 2, 0.000116509, 0
3: MobilenetV1/Logits/Conv2d_1c_1x1/weights_quant/FakeQuantWithMinMaxVars,
1025024, 3, 0.00495183, 67
4: MobilenetV1/MobilenetV1/Conv2d_0/Conv2D_Fold_bias, 128, 2, 0.000161006, 0
5: MobilenetV1/MobilenetV1/Conv2d_0/Relu6, 401408, 3, 0.0235285, 0
6: MobilenetV1/MobilenetV1/Conv2d_0/weights_quant/FakeQuantWithMinMaxVars,
864, 3, 0.0410565, 108
7: MobilenetV1/MobilenetV1/Conv2d_10_depthwise/Relu6, 100352, 3, 0.0235285, 0
8: MobilenetV1/MobilenetV1/Conv2d_10_depthwise/depthwise_Fold_bias, 2048, 2,
0.00039105, 0
9: MobilenetV1/MobilenetV1/Conv2d_10_depthwise/weights_quant/
FakeQuantWithMinMaxVars, 4608, 3, 0.0166203, 131
此處省略中間結果
80: MobilenetV1/MobilenetV1/Conv2d_9_depthwise/depthwise_Fold_bias, 2048, 2,
0.000351091, 0
81: MobilenetV1/MobilenetV1/Conv2d_9_depthwise/weights_quant/
FakeQuantWithMinMaxVars, 4608, 3, 0.014922, 132
```

82: MobilenetV1/MobilenetV1/Conv2d_9_pointwise/Conv2D_Fold_bias, 2048, 2, 0.000161186, 0

83: MobilenetV1/MobilenetV1/Conv2d_9_pointwise/Relu6, 100352, 3, 0.0235285, 0

84: MobilenetV1/MobilenetV1/Conv2d_9_pointwise/weights_quant/ FakeQuantWithMinMaxVars, 262144, 3, 0.00685069, 120

85: MobilenetV1/Predictions/Reshape, 1001, 3, 0.165351, 74

86: MobilenetV1/Predictions/Reshape/shape, 8, 2, 0, 0

87: MobilenetV1/Predictions/Softmax, 1001, 3, 0.00390625, 0

88: Placeholder, 150528, 3, 0.00392157, 0

len: 940650

width, height, channels: 517, 606, 3

input: 88

number of inputs: 1

number of outputs: 1

Interpreter has 90 tensors and 31 nodes

Inputs: 88

Outputs: 87

Tensor   0 MobilenetV1/Logits/AvgPool_1a/AvgPool kTfLiteUInt8 kTfLiteArenaRw   1024 bytes (0.0 MB) 11 11024

Tensor   1 MobilenetV1/Logits/Conv2d_1c_1x1/BiasAdd kTfLiteUInt8 kTfLiteArenaRw   1001 bytes (0.0 MB) 11 11001

Tensor   2 MobilenetV1/Logits/Conv2d_1c_1x1/Conv2D_bias kTfLiteInt32 kTfLiteMmapRo   4004 bytes (0.0 MB) 1001

Tensor   3 MobilenetV1/Logits/Conv2d_1c_1x1/weights_quant/ FakeQuantWithMinMaxVars kTfLiteUInt8 kTfLiteMmapRo   1025024 bytes (1.0 MB) 10011 11024

Tensor   4 MobilenetV1/MobilenetV1/Conv2d_0/Conv2D_Fold_bias kTfLiteInt32 kTfLiteMmapRo   128 bytes (0.0 MB) 32

Tensor   5 MobilenetV1/MobilenetV1/Conv2d_0/Relu6 kTfLiteUInt8 kTfLiteArenaRw   401408 bytes (0.4 MB) 111211232

Tensor   6 MobilenetV1/MobilenetV1/Conv2d_0/weights_quant/ FakeQuantWithMinMaxVars kTfLiteUInt8 kTfLiteMmapRo   864 bytes (0.0 MB)

```
323 33
Tensor   7 MobilenetV1/MobilenetV1/Conv2d_10_depthwise/Relu6 kTfLiteUInt8
kTfLiteArenaRw    100352 bytes (0.1 MB) 11414512
Tensor   8 MobilenetV1/MobilenetV1/Conv2d_10_depthwise/depthwise_
Fold_bias kTfLiteInt32   kTfLiteMmapRo     2048 bytes (0.0 MB) 512
Tensor   9 MobilenetV1/MobilenetV1/Conv2d_10_depthwise/weights_quant/
FakeQuantWithMinMaxVars kTfLiteUInt8   kTfLiteMmapRo    4608 bytes (0.0 MB)
13 3512
Tensor  10 MobilenetV1/MobilenetV1/Conv2d_10_pointwise/Conv2D_Fold_bias
kTfLiteInt32   kTfLiteMmapRo     2048 bytes (0.0 MB) 512
```
此處省略中間結果
```
Tensor  80 MobilenetV1/MobilenetV1/Conv2d_9_depthwise/depthwise_Fold_bias
kTfLiteInt32   kTfLiteMmapRo     2048 bytes (0.0 MB) 512
Tensor  81 MobilenetV1/MobilenetV1/Conv2d_9_depthwise/weights_quant/
FakeQuantWithMinMaxVars kTfLiteUInt8   kTfLiteMmapRo    4608 bytes (0.0 MB)
13 3512
Tensor  82 MobilenetV1/MobilenetV1/Conv2d_9_pointwise/Conv2D_Fold_bias
kTfLiteInt32   kTfLiteMmapRo     2048 bytes (0.0 MB) 512
Tensor  83 MobilenetV1/MobilenetV1/Conv2d_9_pointwise/Relu6 kTfLiteUInt8
kTfLiteArenaRw    100352 bytes (0.1 MB) 11414512
Tensor  84 MobilenetV1/MobilenetV1/Conv2d_9_pointwise/weights_quant/
FakeQuantWithMinMaxVars kTfLiteUInt8   kTfLiteMmapRo   262144 bytes (0.2 MB)
5121 1512
Tensor  85 MobilenetV1/Predictions/Reshape kTfLiteUInt8   kTfLiteArenaRw
1001 bytes ( 0.0 MB) 11001
Tensor  86 MobilenetV1/Predictions/Reshape/shape kTfLiteInt32
kTfLiteMmapRo       8 bytes (0.0 MB) 2
Tensor  87 MobilenetV1/Predictions/Softmax kTfLiteUInt8   kTfLiteArenaRw
1001 bytes (0.0 MB) 11001
Tensor  88 Placeholder         kTfLiteUInt8 kTfLiteArenaRw    150528
bytes (0.1 MB) 12242243
Tensor  89 (null)             kTfLiteUInt8 kTfLiteArenaRw    338688
bytes (0.3 MB) 111211227
```

```
Node    0 Operator Builtin Code    3
  Inputs: 886 4
  Outputs: 5
Node    1 Operator Builtin Code    4
  Inputs: 53332
  Outputs: 31
Node    2 Operator Builtin Code    3
  Inputs: 313634
  Outputs: 35
Node    3 Operator Builtin Code    4
  Inputs: 353938
  Outputs: 37
Node    4 Operator Builtin Code    3
  Inputs: 374240
  Outputs: 41
Node    5 Operator Builtin Code    4
  Inputs: 414544
  Outputs: 43
Node    6 Operator Builtin Code    3
  Inputs: 434846
  Outputs: 47
Node    7 Operator Builtin Code    4
  Inputs: 475150
  Outputs: 49
Node    8 Operator Builtin Code    3
  Inputs: 495452
  Outputs: 53
Node    9 Operator Builtin Code    4
  Inputs: 535756
  Outputs: 55
Node   10 Operator Builtin Code    3
  Inputs: 556058
  Outputs: 59
```

此處省略中間結果

```
Node  26 Operator Builtin Code   3
  Inputs: 253028
  Outputs: 29
Node  27 Operator Builtin Code   1
  Inputs: 29
  Outputs: 0
Node  28 Operator Builtin Code   3
  Inputs: 03 2
  Outputs: 1
Node  29 Operator Builtin Code  22
  Inputs: 186
  Outputs: 85
Node  30 Operator Builtin Code  25
  Inputs: 85
  Outputs: 87
invoked
average time: 203.584 ms
0.666667: 458 bow tie
0.290196: 653 military uniform
0.0117647: 835 suit
0.00784314: 611 jersey
0.00392157: 922 book jacket
```

輸出的內容裡有一些有意義的資訊，例如：

```
# 模型裡包含的張量和節點（node）
Interpreter has 90 tensors and 31 nodes
Inputs: 88
Outputs: 87
# 推測所用的時間
average time: 203.584 ms
```

下面我們透過程式來了解這個應用是怎樣實現的。這個應用只支援 BMP 影像格式，讀取 BMP 是由 read_bmp() 實現的。下面的程式是 read_

bmp() 的一部分，功能是直接從 BMP 檔案裡讀取影像解析度的寬和高。
這段程式的優點是，實現比較簡單，同時能夠避免對協力廠商程式庫的
依賴。

```cpp
const int32_t header_size =
    *(reinterpret_cast<const int32_t*>(img_bytes.data() + 10));
  *width = *(reinterpret_cast<const int32_t*>(img_bytes.data() + 18));
  *height = *(reinterpret_cast<const int32_t*>(img_bytes.data() + 22));
  const int32_t bpp =
    *(reinterpret_cast<const int32_t*>(img_bytes.data() + 28));
```

回 到 label_image.cc， 這 裡 的 void RunInference(Settings* s) 實 現 了
TensorFlow Lite 的推理邏輯。

首先，使用 FlatBuffers 的介面函數從模型的檔案名稱建置一個模型的實
例：

```cpp
std::unique_ptr<tflite::FlatBufferModel> model;
model = tflite::FlatBufferModel::BuildFromFile(s->model_name.c_str());
```

然 後， 呼 叫 TfLiteStatus InterpreterBuilder::operator()() 產 生 一 個 flite::
Interpreter 的實例：

```cpp
std::unique_ptr<tflite::Interpreter> interpreter;
tflite::ops::builtin::BuiltinOpResolver resolver;
tflite::InterpreterBuilder(*model, resolver)(&interpreter);
```

接著，為模型分配記憶體。記憶體管理基本是由 ArenaPlanner 類別來實
現的，程式如下：

```cpp
if (interpreter->AllocateTensors() != kTfLiteOk) {
    LOG(FATAL) << "Failed to allocate tensors!";
}
```

最後，呼叫 Invoke() 執行模型的推理：

```
for (int i = 0; i < s->loop_count; i++) {
  if (interpreter->Invoke() != kTfLiteOk) {
    LOG(FATAL) << "Failed to invoke tflite!\n";
  }}
```

上面的程式組成了使用 TensorFlow Lite 的基本架構，應該還是比較簡單的。

在這個應用裡，有兩個小的特點讀者可以留意一下。

一個特點是使用了 PrintInterpreterState，使用它可以把 Interpreter 的內部狀態列印輸出：

```
if (s->verbose) PrintInterpreterState(interpreter.get());
```

它的實現也比較簡單，基本的邏輯是找到模型的節點和張量，並輸出它們的狀態：

```
for (size_t tensor_index = 0; tensor_index < interpreter->tensors_size();
     tensor_index++) {
  ... ...
}
printf("\n");
for (size_t node_index = 0; node_index < interpreter->nodes_size();
     node_index++) {
  ... ...
}
```

另外一個特點是使用了 Profiler，透過 Profiler 可以把執行狀態列印出來。呼叫 Profiler 的程式如下：

```
profiling::Profiler* profiler = new profiling::Profiler();
```

```
interpreter->SetProfiler(profiler);
if (s->profiling) profiler->StartProfiling();
```

啟動 TFLITE_PROFILING_ENABLED，重新編譯：

```
$ bazel build --cxxopt=-std=c++11 --copt=-DTFLITE_PROFILING_ENABLED
//tensorflow/lite/examples/label_image
```

執行結果如下：

```
$ bazel-bin/tensorflow/lite/examples/label_image/label_image -i ./grace_
hopper.bmp -l ./labels.txt -m ./mobilenet_quant_v1_224.tflite -p 1
Loaded model ./mobilenet_quant_v1_224.tflite
resolved reporter
invoked
average time: 210.206 ms
    13.280, Node    0, OpCode   3, CONV_2D
     7.841, Node    1, OpCode   4, DEPTHWISE_CONV_2D
    11.299, Node    2, OpCode   3, CONV_2D
     3.891, Node    3, OpCode   4, DEPTHWISE_CONV_2D
     8.465, Node    4, OpCode   3, CONV_2D
     7.681, Node    5, OpCode   4, DEPTHWISE_CONV_2D
    15.893, Node    6, OpCode   3, CONV_2D
     1.927, Node    7, OpCode   4, DEPTHWISE_CONV_2D
     7.634, Node    8, OpCode   3, CONV_2D
     3.713, Node    9, OpCode   4, DEPTHWISE_CONV_2D
    14.745, Node   10, OpCode   3, CONV_2D
     0.934, Node   11, OpCode   4, DEPTHWISE_CONV_2D
     7.308, Node   12, OpCode   3, CONV_2D
     1.781, Node   13, OpCode   4, DEPTHWISE_CONV_2D
    14.232, Node   14, OpCode   3, CONV_2D
     1.779, Node   15, OpCode   4, DEPTHWISE_CONV_2D
    14.258, Node   16, OpCode   3, CONV_2D
     1.780, Node   17, OpCode   4, DEPTHWISE_CONV_2D
```

```
    14.243, Node   18, OpCode    3, CONV_2D
     1.779, Node   19, OpCode    4, DEPTHWISE_CONV_2D
    14.269, Node   20, OpCode    3, CONV_2D
     1.778, Node   21, OpCode    4, DEPTHWISE_CONV_2D
    14.242, Node   22, OpCode    3, CONV_2D
     0.462, Node   23, OpCode    4, DEPTHWISE_CONV_2D
     7.633, Node   24, OpCode    3, CONV_2D
     0.822, Node   25, OpCode    4, DEPTHWISE_CONV_2D
    15.075, Node   26, OpCode    3, CONV_2D
     0.032, Node   27, OpCode    1, AVERAGE_POOL_2D
     1.342, Node   28, OpCode    3, CONV_2D
     0.000, Node   29, OpCode   22, RESHAPE
     0.088, Node   30, OpCode   25, SOFTMAX
0.667: 458 bow tie
0.290: 653 military uniform
0.012: 835 suit
0.008: 611 jersey
0.004: 922 book jacket
```

從結果可以發現卷積 CONV_2D 使用的計算時間最多。Profiler 是由 ProfileBuffer 類別實現的，它的基本功能就是記錄各個事件的時間點。由於它是內建的功能，因此使用它的好處是，不需要額外寫程式來實現這些功能。

6.3.2 最小整合（Minimal）

在 ensorflow/lite/examples/ 下的 Minimal 是一個由 GitHub 社群貢獻的小工具。它示範了怎樣讀取模型、建置解譯器，以及執行預測。Minimal 可以作為其他工具的程式基礎，它沒有太複雜的程式，只是提供了 build 檔案的寫法。如果我們建置一個執行檔案，只需使用 f_cc_binary 並取代 srcs 中的原始檔案，非常簡單，程式如下：

```
tf_cc_binary(
    name = "minimal",
    srcs = [
        "minimal.cc",
    ],
    linkopts = tflite_linkopts() + select({
        "//third_party/tensorflow:android": [
            "-pie",  # Android 5.0 and later supports only PIE
            "-lm",   # some builtin ops, e.g., tanh, need -lm
        ],
        "//conditions:default": [],
    }),
    deps = [
        "//third_party/tensorflow/lite:framework",
        "//third_party/tensorflow/lite/kernels:builtin_ops",
    ],
)
```

6.3.3 Graphviz

視覺化是機器學習中了解資料的方法，也是了解程式和內部邏輯的重要
方法。TensorFlow 提供了很多視覺化工具，這裡讓我們嘗試一些有趣的
轉換，例如產生張量流圖。

首先，將 TensorFlow 張量流圖轉為 Graphviz：

```
$ toco --input_file=tf_files/retrained_graph.pb --output_file=tf_files/
retrained.dot --input_format=TENSORFLOW_GRAPHDEF --output_format=GRAPHVIZ_
DOT --input_shape=1,224,224,3 --input_array=input --output_array=final_result
--inference_type=FLOAT --input_data_type=FLOAT
```

Graphviz 產生的模型圖如圖 6-5 所示。

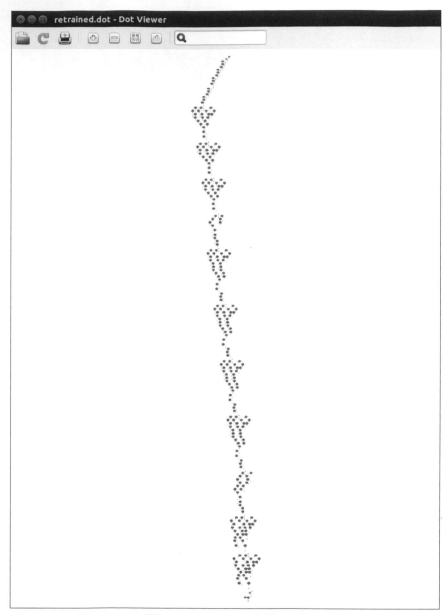

圖 6-5 Graphviz 的模型圖

模型的細節如圖 6-6 所示。

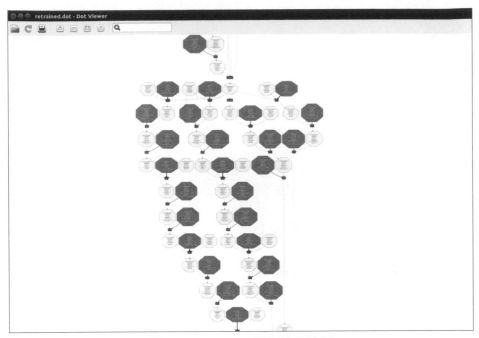

圖 6-6 Graphviz 模型的細節圖

接下來，讓我們將 TensorFlow Lite 模型轉為 Graphviz：

```
$ toco --input_file=tf_files/optimized_graph.lite --output_file=tf_files/
lite.dot --input_format=TFLITE --output_format=GRAPHVIZ_DOT --input_shape=
1,224,224,3 --input_array=input --output_array=final_result --inference_
type=FLOAT --input_data_type=FLOAT
```

由於程式問題，程式不能正常執行，問題在這裡：

```
const auto& input_shape = input_array.shape();
CHECK_EQ(input_shape.dimensions_count(), 4);
```

Toco 需在檢查形狀和形狀尺寸之前檢查緩衝區指標。ooling_util.cc 中 CheckEachArray（cnst Model & model）的基本邏輯是正確的，但缺少邊界檢查，解決方法是增加以下程式：

```
if (!input_array.has_shape()) {
  return;
}
```

在此修復之後，在檢查形狀之前，重新執行下面的指令：

```
$ toco --input_file=tf_files/optimized_graph.lite --output_file=tf_files/
lite.dot --input_format=TFLITE --output_format=GRAPHVIZ_DOT --input_shape=
1,224,224,3 --input_array=input --output_array=final_result --inference_
type= FLOAT --input_data_type=FLOAT
```

獲得如圖 6-7 所示的 TensorFlow Lite 的模型圖。

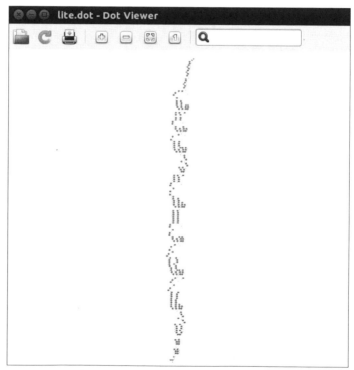

圖 6-7　TensorFlow Lite 的模型圖

TensorFlow Lite 模型放大以後的細節如圖 6-8 所示。

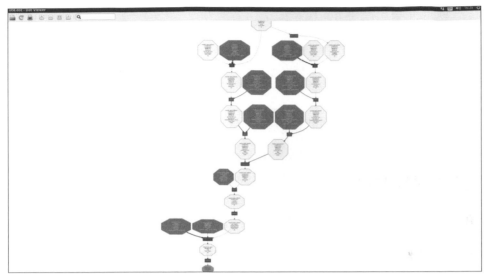

圖 6-8　TensorFlow Lite 模型的細節圖

下面我們把重新訓練的模型產生可視圖，程式如下：

```
$ /opt/tensorflow/bin/toco/toco --input_file=tf_files/retrained_graph.pb
--output_file=tf_files/retrained.dot --input_format=TENSORFLOW_GRAPHDEF
--output_format=GRAPHVIZ_DOT --input_shape=1,224,224,3 --input_array=input
--output_array=final_result --inference_type=FLOAT --input_data_type=FLOAT
```

然後，在 Linux 上，使用 xdot 檢視圖表：

```
xdot tf_files/retrained.dot
```

關於 xdot 的很酷的特性是，你可以點擊一個區塊，這個區塊將被置中，
然後你可以放大和縮小，也可以將 dot 檔案轉為 pdf 檔案。Toco 還提供了
dump_graphviz，可以在轉換時自動產生 dot 檔案，實作方式程式如下：

```
/opt/tensorflow/bin/toco/toco --input_file=tf_files/retrained_graph.pb
--output_file=/tmp/optimized_graph.lite --input_format=TENSORFLOW_GRAPHDEF
--output_format=TFLITE --input_shape=1,224,224,3 --input_array=input
--output_array=final_result --inference_type=FLOAT --input_data_type=FLOAT
--dump_graphviz=/tmp
```

程式執行完畢，產生 dot 檔案如下：

```
-rw-rw-r-- 1 toco_AFTER_ALLOCATION.dot
-rw-rw-r-- 1 toco_AFTER_TRANSFORMATIONS.dot
-rw-rw-r-- 1 toco_AT_IMPORT.dot
```

對於產生的 dot 檔案，讀者可以用 xdot 指令進行檢視。

6.3.4 模型評效

執行下面的指令，產生模型概要。

```
$ summarize_graph --in_graph=tf_files/retrained_graph.pb
```

輸出結果如下：

```
Found 1 possible inputs: (name=Placeholder, type=float(1), shape=
[?,299,299,3])
No variables spotted.
Found 1 possible outputs: (name=final_result, op=Softmax)
Found 21830264 (21.83M) const parameters, 0 (0) variable parameters, and 0
control_edges
Op types used: 490 Const, 378 Identity, 94 Conv2D, 94 FusedBatchNorm,
94 Relu, 15 ConcatV2, 9 AvgPool, 4 MaxPool, 1 Add, 1 Mean, 1 Mul, 1
Placeholder, 1 PlaceholderWithDefault, 1 MatMul, 1 Softmax, 1 Squeeze, 1 Sub
```

執行下面的指令，產生最佳化後的模型概要。

```
$ summarize_graph --in_graph=tensorflow_hub/tf_files/optimized_graph.pb
```

輸出結果如下：

```
Found 1 possible inputs: (name=Placeholder, type=float(1), shape=None)
No variables spotted.
Found 1 possible outputs: (name=final_result, op=Softmax)
```

```
Found 21778616 (21.78M) const parameters, 0 (0) variable parameters, and 0
control_edges
Op types used: 208 Const, 94 BiasAdd, 94 Conv2D, 94 Relu, 15 ConcatV2,
9 AvgPool, 4 MaxPool, 1 Add, 1 MatMul, 1 Mean, 1 Mul, 1 Placeholder, 1
PlaceholderWithDefault, 1 Softmax, 1 Squeeze, 1 Sub
```

可以把模型與 ensorflow/tools/benchmark 一起使用，benchmark_model 的
使用案例如下：

```
benchmark_model --graph=tf_files/retrained_graph.pb --show_flops
--input_ layer=Placeholder --input_layer_type=float
--input_layer_shape=1,299,299,3 --output_layer=final_result
```

輸出結果如下：

```
2018-05-1514:52:05.401180: I tensorflow/core/util/stat_summarizer.cc:468]
=========================== Summary by node type ============================
2018-05-1514:52:05.401186: I tensorflow/core/util/stat_summarizer.cc: 468]
[Node type]  [count]  [avg ms]    [avg %]    [cdf %]   [mem KB][times called]
2018-05-1514:52:05.401195: I tensorflow/core/util/stat_summarizer.cc: 468]
Conv2D       94       69.622      76.439%    76.439% 35869.953        94
2018-05-1514:52:05.401203: I tensorflow/core/util/stat_summarizer.cc: 468]
AvgPool       9        8.866       9.734%    86.173% 8009.600          9
2018-05-1514:52:05.401211: I tensorflow/core/util/stat_summarizer.cc: 468]
BiasAdd      94        6.322       6.941%    93.114%    0.000         94
2018-05-1514:52:05.401220: I tensorflow/core/util/stat_summarizer.cc: 468]
MaxPool       4        2.669       2.930%    96.044% 2834.560          4
2018-05-1514:52:05.401227: I tensorflow/core/util/stat_summarizer.cc: 468]
Relu         63        1.813       1.991%    98.035%    0.000         63
2018-05-1514:52:05.401235: I tensorflow/core/util/stat_summarizer.cc: 468]
ConcatV2     15        1.086       1.192%    99.227% 10678.528        15
2018-05-1514:52:05.401243: I tensorflow/core/util/stat_summarizer.cc: 468]
Const       194        0.376       0.413%    99.640%    0.000        194
2018-05-1514:52:05.401251: I tensorflow/core/util/stat_summarizer.cc: 468]
```

```
Mul         1     0.126    0.138%   99.778%  1072.812      1
2018-05-1514:52:05.401259: I tensorflow/core/util/stat_summarizer.cc: 468]
Mean        1     0.057    0.063%   99.841%    8.192       1
2018-05-1514:52:05.401268: I tensorflow/core/util/stat_summarizer.cc: 468]
Sub         1     0.054    0.059%   99.900%    0.000       1
2018-05-1514:52:05.401276: I tensorflow/core/util/stat_summarizer.cc: 468]
NoOp        1     0.042    0.046%   99.946%    0.000       1
2018-05-1514:52:05.401284: I tensorflow/core/util/stat_summarizer.cc: 468]
MatMul      1     0.015    0.016%   99.963%    0.020       1
2018-05-1514:52:05.401292: I tensorflow/core/util/stat_summarizer.cc: 468]
_Retval     1     0.010    0.011%   99.974%    0.000       1
2018-05-1514:52:05.401300: I tensorflow/core/util/stat_summarizer.cc: 468]
Squeeze     1     0.006    0.007%   99.980%    0.000       1
2018-05-1514:52:05.401309: I tensorflow/core/util/stat_summarizer.cc: 468]
Softmax     1     0.006    0.007%   99.987%    0.000       1
2018-05-1514:52:05.401317: I tensorflow/core/util/stat_summarizer.cc: 468]
Add         1     0.006    0.007%   99.993%    0.000       1
2018-05-1514:52:05.401325: I tensorflow/core/util/stat_summarizer.cc: 468]
_Arg        1     0.004    0.004%   99.998%    0.000       1
2018-05-1514:52:05.401334: I tensorflow/core/util/stat_summarizer.cc: 468]
PlaceholderWithDefault    1    0.002   0.002%  100.000%   0.000  1
2018-05-1514:52:05.401342: I tensorflow/core/util/stat_summarizer.cc: 468]
2018-05-1514:52:05.401347: I tensorflow/core/util/stat_summarizer.cc: 468]
Timings (microseconds): count=217 first=87776 curr=87457 min=84829 max=
207830 avg=91316.4 std=18506
2018-05-1514:52:05.401353: I tensorflow/core/util/stat_summarizer.cc: 468]
Memory (bytes): count=217 curr=58473664(all same)
2018-05-1514:52:05.401359: I tensorflow/core/util/stat_summarizer.cc: 468]
484 nodes observed
2018-05-1514:52:05.401364: I tensorflow/core/util/stat_summarizer.cc: 468]
2018-05-1514:52:06.708874: I tensorflow/tools/benchmark/benchmark_ model.
cc:631] FLOPs estimate: 11.42B
2018-05-1514:52:06.708915: I tensorflow/tools/benchmark/benchmark_ model.
cc:633] FLOPs/second: 260.55B
```

用 TensorFlow Lite
建置機器學習應用

本　章主要介紹開發針對行動端機器學習應用的過程和方法。如果讀者在行動應用程式中使用 TensorFlow Lite 模型，必須選擇預先訓練的模型或自訂的模型，將模型轉為 TensorFLow Lite 的模型格式，再將模型整合到應用程式中。

7.1 模型設計

根據使用場景，讀者可以選擇一種流行的開放原始碼模型，例如 InceptionV3 或 MobileNets，並使用自訂資料集重新訓練這些模型，也可以建置自己的自訂模型。

7.1.1 使用預先訓練的模型

這裡我們先用視覺模型舉例。MobileNets 是 TensorFlow 的以行動優先設計的電腦視覺模型系列中的,這個模型主要在有效地大幅地加強準確性,同時考慮到裝置或嵌入式應用程式的有限資源。

MobileNets 是小型、低延遲、低耗電模型,最佳化參數可以滿足各種資源受限制的應用。這個模型可用於物體的分類、檢測、嵌入和分割,類似其他流行的大型模型,例如 Inception。

Google 為 MobileNets 提供了 16 個經過預先教育訓練的 ImageNet 物體分類檢查點,可用於各種規模的行動應用專案。Inception-v3 是一種影像識別模型,可以實現相當高的準確度,能夠識別 1000 個類別的物體物件,例如「斑馬」和「洗碗機」等。該模型使用卷積神經網路從輸入影像中分析一般特徵,並以具有完全連接和 Softmax 層為基礎的那些特徵對它們進行分類。

「行動端智慧回覆(On Device Smart Reply)」也是一種針對行動裝置的模型,透過建議與上下文相關的資訊,為傳入的文字訊息提供一鍵式回覆。該模型專為記憶體受限裝置(如手錶和手機)而建置,並已成功用於 Android Wear 上的智慧回覆。目前,此模型只支援 Android。

這些預先訓練的模型都可以在 https://www.tensorflow.org/lite/models 上下載。

7.1.2 重新訓練

TensorFlow 也提供了工具和指令稿供開發者使用來進行模型的重新訓練。舉例來說,image_retraining 是一個影像再訓練的工具。

首先，下載原始程式碼：

```
$ git clone https://github.com/tensorflow/hub.git
```

然後，建置執行檔案：

```
$ bazel build tensorflow/examples/image_retraining:retrain
```

TensorFlow Hub 似乎很久沒有人更新和維護了，會出現編譯錯誤。筆者個人的建議是，把需要的原始程式碼複製到 TensorFlow 的倉裡，並做適當的更改。舉例來說，把下面的更改加進 WORKSPACE 裡，就可以滿足建置的依賴要求，實際程式如下：

```
load("@bazel_tools//tools/build_defs/repo:git.bzl", "git_repository")
git_repository(
    name = "protobuf_bzl",
    # v3.6.0
    commit = "ab8edf1dbe2237b4717869eaab11a2998541ad8d",
    remote = "https://github.com/google/protobuf.git",
)
bind(
    name = "protobuf",
    actual = "@protobuf_bzl//:protobuf",
)
bind(
    name = "protobuf_python",
    actual = "@protobuf_bzl//:protobuf_python",
+)
bind(
    name = "protobuf_python_genproto",
    actual = "@protobuf_bzl//:protobuf_python_genproto",
)
bind(
    name = "protoc",
```

```
    actual = "@protobuf_bzl//:protoc",
)
# Using protobuf version 3.6.0
http_archive(
    name = "com_google_protobuf",
    strip_prefix = "protobuf-3.6.0",
    urls = ["https://github.com/google/protobuf/archive/v3.6.0.zip"],
)
```

上述程式編譯後，就獲得可執行檔。在 retrain.py 裡可以看到如下所示的
程式，功能是把模型的參數傳給執行檔案。

```
parser.add_argument(
    '--tfhub_module',
    type=str,
    default=(
        'https://tfhub.dev/google/imagenet/inception_v3/feature_ vector/1'),
        help="""
        Which TensorFlow Hub module to use.
        See https://github.com/tensorflow/hub/blob/master/docs/modules/image.md
        for some publicly available ones.
    """)
```

準備好資料檔案後，就可以進行影像的再訓練。首先，把圖片儲存到一
個資料夾（這裡我們把影像儲存到 flower_photos）中。然後，執行指令
稿，它的中間輸出將在系統 mp 資料夾中的 fhub_modules 中。下面的
指令稿分別對三個模型進行再訓練，這三個模型分別是 inception_v3、
mobilenet 和 mobilenet 的定點數模型。

```
$ bazel-bin/examples/image_retraining/retrain --image_dir flower_photos
--tfhub_module https://tfhub.dev/google/imagenet/inception_v3/feature_vector/1
--saved_model_dir inception_v3/saved_model/ --output_graph inception_v3/
retain_graph.pb --output_labels inception_v3/label.txt --summaries_dir
```

```
inception_v3/summaries/ --bottleneck_dir inception_v3/bottleneck/

$ bazel-bin/examples/image_retraining/retrain --image_dir flower_photos
--tfhub_module https://tfhub.dev/google/imagenet/mobilenet_v1_100_224/
feature_ vector/1 --saved_model_dir mobilenet_float/saved_model/
--output_graph mobilenet_float/retain_graph.pb --output_labels
mobilenet_float/label.txt --summaries_dir mobilenet_float/summaries/
--bottleneck_dir mobilenet_float/ bottleneck/

$ bazel-bin/examples/image_retraining/retrain --image_dir flower_photos
--tfhub_module https://tfhub.dev/google/imagenet/mobilenet_v1_100_224/
quantops/feature_vector/1  --saved_model_dir mobilenet_quant/saved_model/
--output_graph mobilenet_quant/retain_graph.pb --output_labels
mobilenet_ quant/label.txt --summaries_dir mobilenet_quant/summaries/
--bottleneck_dir mobilenet_quant/bottleneck/
```

7.1.3 使用瓶頸（Bottleneck）

根據機器速度的不同，上面指令稿可能需要很長時間才能完成訓練。執行的第一個階段是分析儲存的所有影像，並計算和快取每個映射的瓶頸值。「瓶頸」是一個非正式術語，實作中我們經常使用它，它表示在最後輸出層之前的層（在 TensorFlow Hub 裡將其稱為「影像特徵向量」）。此倒數第二層經過訓練，已經可以滿足區分的要求，進而輸出一組足夠好的結果。這表示它必須包含一個有意義且緊湊的影像概要，即它必須包含足夠的資訊，以便分類器在一組非常有限的值中做出正確的選擇。我們把最後一層經過再訓練就可以用於新類別的原因就在於此。結果表明，用於區分 ImageNet 中所有 1000 個類別所需的資訊通常也可用於區分新的類型。

因為每個影像在訓練期間多次重複使用，並且計算每個瓶頸耗時很長，如果我們能把瓶頸的值快取到磁碟上，就可以節省大量時間。預設情況下，它們儲存在 /tmp/bottleneck 目錄中，如果重新執行指令稿，它們可以被重用。使用瓶頸進行訓練的實現程式如下：

```
IMAGE_SIZE=224
ARCHITECTURE="mobilenet_0.50_${IMAGE_SIZE}"
$ bazel-bin/examples/image_retraining/retrain --bottleneck_dir=tf_files/
bottlenecks --how_many_training_steps=500 --model_dir=tf_files/models/
--summaries_dir=tf_files/training_summaries/"${ARCHITECTURE}"
--output_graph=tf_files/retrained_graph.pb --output_labels=tf_files/
retrained_labels.txt --architecture="${ARCHITECTURE}" --image_dir=
flower_photos
```

我們現在可以使用訓練好的模型，嘗試去檢測一張影像，實作方式程式如下：

```
$ label_image --graph=tf_files/retrained_graph.pb --image=flower_photos/
sunflowers/24459548_27a783feda.jpg --input_layer=Placeholder
--output_layer= final_result --labels=tf_files/retrained_labels.txt
--input_width=299 --input_height=299
```

執行結果如下：

```
2018-05-1015:53:16.354204: I tensorflow/core/platform/cpu_feature_guard.
cc:141] Your CPU supports instructions that this TensorFlow binary was not
compiled to use: SSE4.1 SSE4.2 AVX AVX2 FMA
2018-05-1015:53:17.986350: I tensorflow/examples/label_image/main.cc:251]
sunflowers (3): 0.825468
2018-05-1015:53:17.986390: I tensorflow/examples/label_image/main.cc:251]
tulips (4): 0.0628504
2018-05-1015:53:17.986401: I tensorflow/examples/label_image/main.cc:251]
daisy (0): 0.0562632
2018-05-1015:53:17.986408: I tensorflow/examples/label_image/main.cc:251]
dandelion (1): 0.0406035
2018-05-1015:53:17.986414: I tensorflow/examples/label_image/main.cc:251]
roses (2): 0.0148145
```

現在，讀者還可以使用最佳化器，對產生的模型進行進一步的最佳化：

```
$ bazel build tensorflow/python/tools:optimize_for_inference
$ bazel-bin/tensorflow/python/tools/optimize_for_inference
--input=tf_files/retrained_graph.pb --output=tf_files/optimized_graph.pb
--input_name= "Placeholder" --output_name="final_result"
```

最佳化的效果如下：

```
2018-05-1016:16:26.601517: I tensorflow/lite/toco/graph_transformations/
graph_transformations.cc:39] Before Removing unused ops: 1072 operators,
1658 arrays (0 quantized)
2018-05-1016:16:26.631377: I tensorflow/lite/toco/graph_transformations/
graph_transformations.cc:39] Before general graph transformations: 1072
operators, 1658 arrays (0 quantized)
2018-05-1016:16:26.743531: I tensorflow/lite/toco/graph_transformations/
graph_transformations.cc:39] After general graph transformations pass 1: 128
operators, 323 arrays (0 quantized)
2018-05-1016:16:26.745757: I tensorflow/lite/toco/graph_transformations/
graph_transformations.cc:39] After general graph transformations pass 2: 126
operators, 319 arrays (0 quantized)
2018-05-1016:16:26.747896: I tensorflow/lite/toco/graph_transformations/
graph_transformations.cc:39] Before dequantization graph transformations:
126 operators, 319 arrays (0 quantized)
2018-05-1016:16:26.749797: I tensorflow/lite/toco/allocate_transient_
arrays.cc:329] Total transient array allocated size: 0 bytes, theoretical
optimal value: 0 bytes.
```

現在，我們可以檢驗一下訓練過的模型，進而確定效果。這裡我們使用 label_image 工具：

```
$ label_image --graph=tf_files/optimized_graph.pb --image=flower_photos/
sunflowers/24459548_27a783feda.jpg --input_layer=Placeholder
--output_layer= final_result --labels=tf_files/retrained_labels.txt
--input_width=299 --input_height=299
```

輸出結果如下：

```
2018-05-1016:11:33.589364: I tensorflow/core/platform/cpu_feature_guard.
cc:141] Your CPU supports instructions that this TensorFlow binary was not
compiled to use: SSE4.1 SSE4.2 AVX AVX2 FMA
2018-05-1016:11:35.120186: I tensorflow/examples/label_image/main.cc:251]
sunflowers (3): 0.825468
2018-05-1016:11:35.120239: I tensorflow/examples/label_image/main.cc:251]
tulips (4): 0.0628507
2018-05-1016:11:35.120263: I tensorflow/examples/label_image/main.cc:251]
daisy (0): 0.0562634
2018-05-1016:11:35.120269: I tensorflow/examples/label_image/main.cc:251]
dandelion (1): 0.0406035
2018-05-1016:11:35.120276: I tensorflow/examples/label_image/main.cc:251]
roses (2): 0.0148145
```

確認結果之後，就可以使用 Toco 工具把訓練好的模型轉換成 TensorFlow
Lite 的模型。如果你的預先安裝 Toco 有以下問題，你可以重新編譯一
下。

```
$ toco
TOCO from pip install is currently not working on command line.
Please use the python TOCO API or use
bazel run tensorflow/lite:toco -- <args> from a TensorFlow source dir.
```

執行下面的指令，就可以重新編譯 Toco。

```
$ bazel build tensorflow/lite/toco:toco
```

然後我們就可以使用 Toco 了。執行下面的指令，把 inception_v3 和
mobilenet 的模型轉換成 TensorFlow Lite 的模型。

```
$ /opt/tensorflow/bin/toco/toco --input_file=inception_v3/retain_graph.pb
--output_file=inception_v3/optimized_graph.lite --input_format=TENSORFLOW_
```

```
GRAPHDEF --output_format=TFLITE --input_shape=1,224, 224,3 --input_array=input
--output_array=final_result --inference_type=FLOAT --input_data_type=FLOAT
--dump_graphviz=inception_v3/

$ /opt/tensorflow/bin/toco/toco --input_file=mobilenet_float/retain_ graph.pb
--output_file=mobilenet_float/optimized_graph.lite --input_format=
TENSORFLOW_GRAPHDEF --output_format=TFLITE --input_shape=1,224,224,3
--input_ array=input --output_array=final_result --inference_type=FLOAT
--input_data_ type=FLOAT --dump_graphviz=mobilenet_float/
```

上面我們把預先訓練的模型在自己的資料集上進行訓練,原始的模型可以對 1000 個類別進行分類。如果這些類別不足以滿足你的使用需求,則需要重新訓練模型。這種技術稱為遷移學習,從已經訓練過問題的模型開始,在類似的問題上重新訓練模型。從頭開始深度學習可能需要數天時間,但遷移學習相當快。為了進行遷移學習,讀者需要產生標有相關類別的自訂資料集。

開發人員也可以選擇使用 TensorFlow 先訓練自訂模型。即先撰寫模型,再對資料進行訓練。

使用 TensorFlow Lite 在行動端上的開發和使用 TensorFlow Mobile 非常類似。一般來講,應用程式的程式重用度非常高,要做的事基本就是,根據模型的不同對模型的輸入和輸出做出調整。

前面,我們對如何建置應用做了很多說明,這裡就不重複了。TensorFlow Lite 支援 TensorFlow 運算子的子集。有關支援的運算子及其用法,可以參閱 TensorFlow 的官方文件。TensorFlow Lite 也意識到這個問題,Google 正在進行開發,同時也在研究可以快速使用已有運算元的方法。

7.2 開發應用

這裡我們介紹開發 TensorFlow Lite 應用所使用的介面和一些技術要點。

7.2.1 程式介面

TensorFlow Lite 為開發人員提供了非常容易使用的 API 介面，只需幾步驟就可以完成一個簡單應用的開發。下面是範例程式：

```java
// 定義 TfLite
private Interpreter tfLite;

// Tflite，建立實例
try {
  tfLite = new Interpreter(loadModelFile(MODEL_FILE));
} catch (IOException e) {
  Log.e(TAG, "Failed to create TFLite");
  finish();
}

// 執行 Tflite
tfLite.run(data, label);
```

只需簡單的三步驟就可以執行應用。步驟一，定義一個解譯器；步驟二，產生一個實例；步驟三，執行。

TensorFlow Lite 提供了 6 個建置函數可以產生一個解譯器。舉例來説，應用可以直接傳入一個檔案的物件：

```java
/**
 * Initializes a {@code Interpreter}
 *
 * @param modelFile: a File of a pre-trained TF Lite model.
 */
```

```
public Interpreter(@NonNull File modelFile) {
  this(modelFile, /*options = */ null);
}
```

筆者推薦讀者使用下面的 API 去產生解譯器：

```
/**
 * Initializes a {@code Interpreter} with a {@code ByteBuffer} of a model
file and a set of custom
 * {@link #Options}.
 *
 * <p>The ByteBuffer should not be modified after the construction of a
{@code Interpreter}. The
 * {@code ByteBuffer} can be either a {@code MappedByteBuffer} that memory-
maps a model file, or a
 * direct {@code ByteBuffer} of nativeOrder() that contains the bytes
content of a model.
 */
public Interpreter(@NonNull ByteBuffer byteBuffer, Options options) {
  wrapper = new NativeInterpreterWrapper(byteBuffer, options);
}
```

這種產生解譯器的方法的優點是，可以減少模型載入時間。建議使用記憶體共用，先產生 MappedByteBuffer，再傳入解析器。

對於執行函數，有兩個選擇，即 run 介面和 runForMultipleInputsOutputs 介面。這兩個介面區別不大。如果模型只有一個輸入，就用 run 介面；如果模型有多個輸入，就用 runForMultipleInputsOutputs 介面。

介面 run 的程式如下：

```
/**
 * Runs model inference if the model takes only one input, and provides only
one output.
```

```
   *
   * <p>Warning: The API runs much faster if {@link ByteBuffer} is used as
input data type. Please
   * consider using {@link ByteBuffer} to feed primitive input data for better
performance.
   *
   * @param input an array or multidimensional array, or a {@link ByteBuffer}
of primitive types
   *       including int, float, long, and byte. {@link ByteBuffer} is the
preferred way to pass large
   *       input data for primitive types, whereas string types require using
the (multi-dimensional)
   *       array input path. When {@link ByteBuffer} is used, its content should
remain unchanged
   *       until model inference is done. A {@code null} value is allowed only
if the caller is using
   *       a {@link Delegate} that allows buffer handle interop, and such a
buffer has been bound to
   *       the input {@link Tensor}.
   * @param output a multidimensional array of output data, or a {@link
ByteBuffer} of primitive
   *       types including int, float, long, and byte. A null value is allowed
only if the caller is
   *       using a {@link Delegate} that allows buffer handle interop, and such
a buffer has been
   *       bound to the output {@link Tensor}. See also {@link Options#setAllowB
ufferHandleOutput()}.
   */
public void run(Object input, Object output) {
  Object[] inputs = {input};
  Map<Integer, Object> outputs = new HashMap<>();
  outputs.put(0, output);
  runForMultipleInputsOutputs(inputs, outputs);
}
```

介面 runForMultipleInputsOutputs 的程式如下：

```
/**
/**
 * Runs model inference if the model takes multiple inputs, or returns
multiple outputs.
 *
 * <p>Warning: The API runs much faster if {@link ByteBuffer} is used as
input data type. Please
 * consider using {@link ByteBuffer} to feed primitive input data for better
performance.
 *
 * <p>Note: {@code null} values for invididual elements of {@code inputs}
and {@code outputs} is
 * allowed only if the caller is using a {@link Delegate} that allows buffer
handle interop, and
 * such a buffer has been bound to the corresponding input or output {@link
Tensor}(s).
 *
 * @param inputs an array of input data. The inputs should be in the same
order as inputs of the
 *     model. Each input can be an array or multidimensional array, or a
{@link ByteBuffer} of
 *     primitive types including int, float, long, and byte. {@link
ByteBuffer} is the preferred
 *     way to pass large input data, whereas string types require using the
(multi-dimensional)
 *     array input path. When {@link ByteBuffer} is used, its content should
remain unchanged
 *     until model inference is done.
 * @param outputs a map mapping output indices to multidimensional arrays of
output data or {@link
 *     ByteBuffer}s of primitive types including int, float, long, and byte.
It only needs to keep
```

```
*      entries for the outputs to be used.
*/
public void runForMultipleInputsOutputs(
    @NonNull Object[] inputs, @NonNull Map<Integer, Object> outputs) {
  checkNotClosed();
  wrapper.run(inputs, outputs);
}
```

下面是一個輸入執行函數對應的原生 C 的實現：

```
public void run(@NonNull Object input, @NonNull Object output) {
  Object[] inputs = {input};
  Map<Integer, Object> outputs = new HashMap<>();
  outputs.put(0, output);
  runForMultipleInputsOutputs(inputs, outputs);
}
```

下面是多個輸入執行函數對應的原生 C 程式的實現，讀者可以參考。

```
public void runForMultipleInputsOutputs(
    @NonNull Object[] inputs, @NonNull Map<Integer, Object> outputs) {
  if (wrapper == null) {
    throw new IllegalStateException("Internal error: The Interpreter has
already been closed.");
  }
  Tensor[] tensors = wrapper.run(inputs);
  if (outputs == null || tensors == null || outputs.size() > tensors.length) {
    throw new IllegalArgumentException("Output error: Outputs do not match
with model outputs.");
  }
  final int size = tensors.length;
  for (Integer idx : outputs.keySet()) {
  if (idx == null || idx < 0 || idx >= size) {
    throw new IllegalArgumentException(
```

```
String.format(
    "Output error: Invalid index of output %d (should be in range [0, %d))",
    idx, size));
}
tensors[idx].copyTo(outputs.get(idx));
}
}
```

7.2.2 執行緒和效能

TensorFlow Lite 是單執行緒的執行程式。它會根據模型循序執行。但是在 Interpreter.java 裡，TensorFlow Lite 提供了設定執行緒的介面，現在來看一下它的內部實現：

```
interpreter->SetNumThreads(static_cast<int>(num_threads));
```

在 C++ 的實現過程中，程式把執行緒數傳給了 eigen。現在，我們看到執行緒數是針對 eigen 的，不是針對 TensorFlow Lite。實際程式如下：

```
void Interpreter::SetNumThreads(int num_threads) {
  context_.recommended_num_threads = num_threads;

  gcmm_support::SetNumThreads(&context_, num_threads);
  eigen_support::SetNumThreads(&context_, num_threads);
}
```

下面的程式是解譯器執行過程的實現，可以看到這是一個單執行緒的執行：

```
for (int execution_plan_index = 0;
    execution_plan_index < execution_plan_.size(); execution_plan_index++) {
  if (execution_plan_index == next_execution_plan_index_to_prepare_) {
    TF_LITE_ENSURE_STATUS(PrepareOpsAndTensors());
```

```
    TF_LITE_ENSURE(&context_, next_execution_plan_index_to_prepare_ >=
        execution_plan_index);
    }
    int node_index = execution_plan_[execution_plan_index];
    TfLiteNode& node = nodes_and_registration_[node_index].first;
    const TfLiteRegistration& registration =
        nodes_and_registration_[node_index].second;
    ** __android_log_print(ANDROID_LOG_VERBOSE, "TFLITE", "node:%d",
registration.builtin_code);**
    SCOPED_OPERATOR_PROFILE(profiler_, node_index);
```

現在，來看一下多執行緒對執行的影響，我們可以使用開放原始碼的影像檢測器做測試，分別使用 1 個執行緒和 10 個執行緒，測試結果如下。

1 個執行緒：

App	CPU float	CPU quant	NPU
image classifier / frame	800ms	150ms	70ms

10 個執行緒：

App	CPU float	CPU quant	NPU
image classifier / frame	800ms	150ms	70ms

看上去好像差別並不大。從其他測試來看，對有些應用的效能提升不是很大。注意，這裡的多執行緒主要是軟體的實現。

7.2.3 模型最佳化

在行動裝置上執行機器學習應用的關鍵是最佳化模型。有很多方法可以最佳化模型，例如定點化、模型剪裁、平行化處理等。在 TensorFlow 出來之前，就已經有了為行動裝置等小型運算裝置最佳化的模型，我們在學習 TensorFlow 怎樣為行動裝置設計之外，還要檢查一下這些模型的內部結構，看看它們的實際效果，並為後面的模型最佳化做準備。

1 量子化模型

我們先來看一下 Mobile Net，這是一個為行動裝置等做最佳化的模型。這次，我們不用直接下載的方法，而直接從 Android 的應用中檢驗模型。透過這個實例，我也希望讀者可以思考怎樣保護自己的模型。

首先，編譯 TensorFlow Lite 的應用：

```
$ bazel build -c opt --cxxopt=---std=c++11 --fat_apk_cpu=arm64-v8a
//tensorflow/lite/java/demo/app/src/main:TfLiteCameraDemo
```

假設我們從網上下載了這個應用，那麼我們可以先解壓 APK：

```
$ unzip TfLiteCameraDemo.apk
```

然後，透過檢查應用的檔案，找到模型檔案：

```
$ ls -l assets/
total 4220
-rw-rw-rw- 1  BUILD
-rw-rw-rw- 1  labels_imagenet_slim.txt
-rw-rw-rw- 1  labels_mobilenet_quant_v1_224.txt
-rw-rw-rw- 1  labels.txt
-rw-rw-rw- 1  mobilenet_quant_v1_224.tflite
-rw-rw-rw- 1  WORKSPACE
```

接著，使用 Toco 工具將其轉為 Graphviz 檔案：

```
$ toco --input_file=mobilenet_quant_v1_224.tflite --output_file=mobilenet_
quant_v1_224.dot --input_format=TFLITE  --output_format=GRAPHVIZ_DOT
```

執行上面的指令後獲得了如圖 7-1 所示的 Mobile Net 模型圖。

圖 7-1　Mobile Net 模型圖

最後，在 Camera2BasicFragment.java 中按照上面的範例程式啟用呼叫記
錄檔記錄，重新建置並執行它，我們應該在 Android logcat 中看到有類
似的內容：

```
invoke start
node:3 kTfLiteBuiltinConv2d
node:4 kTfLiteBuiltinDepthwiseConv2d
node:3 kTfLiteBuiltinConv2d
node:4 kTfLiteBuiltinDepthwiseConv2d
node:3 kTfLiteBuiltinConv2d
```

```
node:4 kTfLiteBuiltinDepthwiseConv2d
node:3 kTfLiteBuiltinConv2d
node:4 kTfLiteBuiltinDepthwiseConv2d
node:3 kTfLiteBuiltinConv2d
node:4 kTfLiteBuiltinDepthwiseConv2d
node:3 kTfLiteBuiltinConv2d
node:4 kTfLiteBuiltinDepthwiseConv2d
node:3 kTfLiteBuiltinConv2d
node:4 kTfLiteBuiltinDepthwiseConv2d
node:3 kTfLiteBuiltinConv2d
node:4 kTfLiteBuiltinDepthwiseConv2d
node:3 kTfLiteBuiltinConv2d
node:4 kTfLiteBuiltinDepthwiseConv2d
node:3 kTfLiteBuiltinConv2d
node:4 kTfLiteBuiltinDepthwiseConv2d
node:3 kTfLiteBuiltinConv2d
node:4 kTfLiteBuiltinDepthwiseConv2d
node:3 kTfLiteBuiltinConv2d
node:4 kTfLiteBuiltinDepthwiseConv2d
node:3 kTfLiteBuiltinConv2d
node:4 kTfLiteBuiltinDepthwiseConv2d
node:3 kTfLiteBuiltinConv2d
node:1 kTfLiteBuiltinAveragePool2d
node:3 kTfLiteBuiltinConv2d
node:22 kTfLiteBuiltinReshape
node:25 kTfLiteBuiltinSoftmax
invoke end
```

一共 31 個 Ops，這些 Ops 是在 builtin_ops.h 中定義的，現在我們可以做
交換參考，可以用 logcat 直接輸出 Ops 的名稱。在上面的輸出中，基本
都是 kTfLiteBuiltinConv2d 和 kTfLiteBuiltinDepthwiseConv2d。從結構圖
中我們也可以看到這個模型是非常簡單和高效的。

2 浮點數模型

下面讓我們來看一下 Inception v3 模型的內在結構,讀者可以使用上面的方法,也可以從網站下載:

https://storage.googleapis.com/download.tensorflow.org/models/tflite/inception_v3_slim_2016_android_2017_11_10.zip

下載後解壓縮,獲得以下檔案:

```
inflating: inceptionv3_slim_2016.tflite
inflating: imagenet_slim_labels.txt
```

我們使用 Toco,用同樣的方法,可以獲得模型的視圖,程式如下:

```
toco --input_file=tensorflow/lite/java/demo/app/src/main/assets/
inceptionv3_slim_2016.tflite --output_file=lite.dot --input_format=TFLITE
--output_format=GRAPHVIZ_DOT
```

Inceptionv3 模型如圖 7-2 所示。

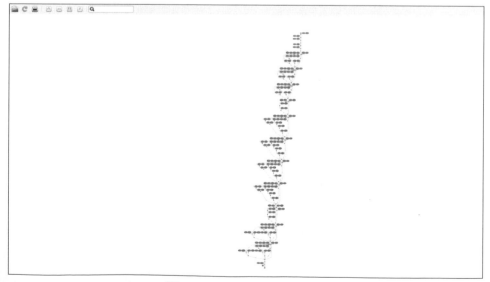

圖 7-2 Inceptionv3 模型圖

這個模型圖（圖 7-2）比 MobileNet 複雜得多。讓我們放大這個圖的一部分，看看解譯器如何執行圖。在圖 7-2 中，單一輸入是 MaxPool 的輸出，單一輸出是啟動。需要連接 4 個輸入，一共包含 6 個 conv2d。從圖 7-2 中，我們可以很容易地發現 6 個 conv2d 操作可以並存執行。但是，從後面的 logcat 可以看出，interperter 是循序執行它們的。模型放大後的細節如圖 7-3 所示。

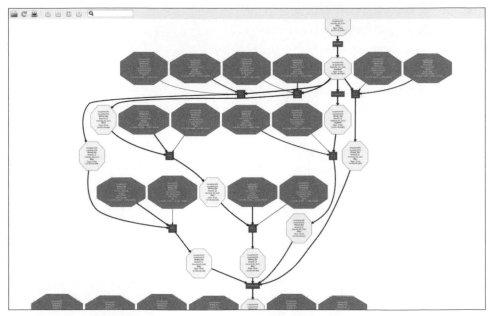

圖 7-3　Inceptionv3 模型放大圖

同樣，按照上面的方法啟用及呼叫記錄檔記錄，並執行它，我們應該在 logcat 中有類似的內容：

```
invoke start
node:3    kTfLiteBuiltinConv2d
node:3    kTfLiteBuiltinConv2d
node:3    kTfLiteBuiltinConv2d
node:17   kTfLiteBuiltinMaxPool2d
```

```
node:3  kTfLiteBuiltinConv2d
node:17 kTfLiteBuiltinMaxPool2d
node:3  kTfLiteBuiltinConv2d
node:3  kTfLiteBuiltinConv2d
node:3  kTfLiteBuiltinConv2d
node:3  kTfLiteBuiltinConv2d
node:3  kTfLiteBuiltinConv2d
node:1  kTfLiteBuiltinAveragePool2d
node:3  kTfLiteBuiltinConv2d
node:2  kTfLiteBuiltinConcatenation
node:3  kTfLiteBuiltinConv2d
node:3  kTfLiteBuiltinConv2d
node:1  kTfLiteBuiltinAveragePool2d
node:3  kTfLiteBuiltinConv2d
node:2  kTfLiteBuiltinConcatenation
node:3  kTfLiteBuiltinConv2d
node:3  kTfLiteBuiltinConv2d
node:3  kTfLiteBuiltinConv2d
node:3  kTfLiteBuiltinConv2d
node:1  kTfLiteBuiltinAveragePool2d
node:3  kTfLiteBuiltinConv2d
node:2  kTfLiteBuiltinConcatenation
node:3  kTfLiteBuiltinConv2d
node:3  kTfLiteBuiltinConv2d
node:3  kTfLiteBuiltinConv2d
node:3  kTfLiteBuiltinConv2d
node:17 kTfLiteBuiltinMaxPool2d
node:2  kTfLiteBuiltinConcatenation
node:3  kTfLiteBuiltinConv2d
node:3  kTfLiteBuiltinConv2d
node:3  kTfLiteBuiltinConv2d
node:3  kTfLiteBuiltinConv2d
node:3  kTfLiteBuiltinConv2d
node:3  kTfLiteBuiltinConv2d
```

```
node:1   kTfLiteBuiltinAveragePool2d
node:3   kTfLiteBuiltinConv2d
node:2   kTfLiteBuiltinConcatenation
node:3   kTfLiteBuiltinConv2d
node:3   kTfLiteBuiltinConv2d
node:3   kTfLiteBuiltinConv2d
node:3   kTfLiteBuiltinConv2d
node:3   kTfLiteBuiltinConv2d
node:3   kTfLiteBuiltinConv2d
node:3   kTfLiteBuiltinConv2d
node:1   kTfLiteBuiltinAveragePool2d
node:3   kTfLiteBuiltinConv2d
node:2   kTfLiteBuiltinConcatenation
node:3   kTfLiteBuiltinConv2d
node:3   kTfLiteBuiltinConv2d
node:1   kTfLiteBuiltinAveragePool2d
node:3   kTfLiteBuiltinConv2d
node:2   kTfLiteBuiltinConcatenation
node:3   kTfLiteBuiltinConv2d
node:3   kTfLiteBuiltinConv2d
node:3   kTfLiteBuiltinConv2d
node:3   kTfLiteBuiltinConv2d
node:3   kTfLiteBuiltinConv2d
node:1   kTfLiteBuiltinAveragePool2d
node:3   kTfLiteBuiltinConv2d
node:2   kTfLiteBuiltinConcatenation
node:3   kTfLiteBuiltinConv2d
node:3   kTfLiteBuiltinConv2d
node:17  kTfLiteBuiltinMaxPool2d
node:2   kTfLiteBuiltinConcatenation
node:3   kTfLiteBuiltinConv2d
node:3   kTfLiteBuiltinConv2d
node:2   kTfLiteBuiltinConcatenation
node:3   kTfLiteBuiltinConv2d
```

```
node:3  kTfLiteBuiltinConv2d
node:2  kTfLiteBuiltinConcatenation
node:1  kTfLiteBuiltinAveragePool2d
node:3  kTfLiteBuiltinConv2d
node:2  kTfLiteBuiltinConcatenation
node:3  kTfLiteBuiltinConv2d
node:3  kTfLiteBuiltinConv2d
node:2  kTfLiteBuiltinConcatenation
node:3  kTfLiteBuiltinConv2d
node:3  kTfLiteBuiltinConv2d
node:3  kTfLiteBuiltinConv2d
node:3  kTfLiteBuiltinConv2d
node:2  kTfLiteBuiltinConcatenation
node:1  kTfLiteBuiltinAveragePool2d
node:3  kTfLiteBuiltinConv2d
node:2  kTfLiteBuiltinConcatenation
node:1  kTfLiteBuiltinAveragePool2d
node:3  kTfLiteBuiltinConv2d
node:22 kTfLiteBuiltinReshape
```

可以看到一共有 61 個運算，所以這個模型比 MobileNet 複雜得多。

另外一種方法，我們還可以使用 Android Trace 來視覺化執行時期的資訊。在 Interpreter.java 中，我們進行了以下修改：

```java
public void run(@NonNull Object input, @NonNull Object output) {
  Trace.beginSection("run");
  Object[] inputs = {input};
  Map<Integer, Object> outputs = new HashMap<>();
  outputs.put(0, output);
  runForMultipleInputsOutputs(inputs, outputs);
  Trace.endSection();
}
```

執行應用後，可以獲得 Trace 檔案。再使用下面的指令，把結果轉換成
網頁。

```
python /opt/Android/sdk/platform-tools/systrace/systrace.py --app com.
example.android.tflitecamerademo -t 5 -o result.html
```

最後，載入 result.html 到瀏覽器中，獲得的結果如圖 7-4 所示。

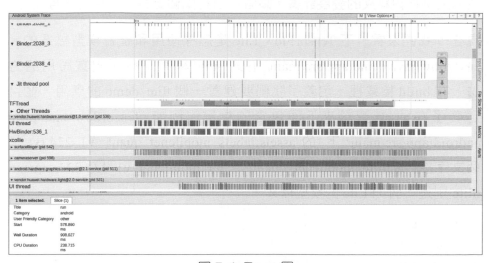

圖 7-4　Trace 圖

從圖 7-4 中讀者可以進一步分析執行的效能和最佳化點。

7.3 TensorFlow Lite 的應用

TensorFlow Lite 提供了幾個應用供讀者參考，這幾個應用也可以作為開
發人員開發產品的基礎。有意思的是，TensorFlow Lite 的應用實例儲存
在兩個資料夾中，一個資料夾是 ensorflow/lite/java/ 下的 demo，另一個
是 ensorflow/lite/examples/ 下的 android。

tensorflow/lite/examples/ 下的 android 和 TensorFlow Mobile 的應用實例有很深的淵源，ensorflow/lite/java/ 下的 demo 一開始是作為 Android 的 Java 的應用實例出現的。兩個資料夾裡的活躍度都不低，現在還在保持更新。

另外，程式的相似度很高，使用的模型也很相似。現在還是不太清楚以後的這兩個資料夾的發展計畫。讀者可以把兩個資料夾都作為參考。

現在，我們來看一下 TensorFlow Lite 是怎樣使用和整合模型檔案的。ensorflow/java/demo/app/src/main/ 下的 build 檔案定義了建置應用的方法，在 build 檔案裡，定義了我們要建置一個 flite_demo 的應用，並且定義了在 assets 資料夾下面包含的檔案。例如 //tensorflow/examples/android/app/src/main/ 下 的 assets:labels_mobilenet_quant_ v1_224.txt 是個文字檔，@tflite_mobilenet_quant// 下的 :mobilenet_v1_1.0_224_ quant.tflite 則是一個 TFLite 的模型檔案。

```
android_binary(
    name = "tflite_demo",
    srcs = glob([
        "app/src/main/java/**/*.java",
    ]),
    aapt_version = "aapt",
    assets = [
        "//tensorflow/lite/examples/android/app/src/main/assets:labels_
mobilenet_quant_v1_224.txt",
        "@tflite_mobilenet_quant//:mobilenet_v1_1.0_224_quant.tflite",
        "@tflite_conv_actions_frozen//:conv_actions_frozen.tflite",
        "//tensorflow/lite/examples/android/app/src/main/assets:conv_
actions_labels.txt",
        "@tflite_mobilenet_ssd//:mobilenet_ssd.tflite",
        "@tflite_mobilenet_ssd_quant//:detect.tflite",
        "//tensorflow/lite/examples/android/app/src/main/assets:box_ priors.
```

```
txt",
        "//tensorflow/lite/examples/android/app/src/main/assets:coco_
labels_list.txt",
    ],
```

關於 mobilenet_v1_1.0_224_quant.tflite，在 ensorflow/workspace.bzl 裡，
可以看到這段程式：

```
tf_http_archive(
    name = "tflite_mobilenet_quant",
    build_file = clean_dep("//third_party:tflite_mobilenet_quant.BUILD"),
    sha256 = "d32432d28673a936b2d6281ab0600c71cf7226dfe4cdcef3012555
f691744166",
    urls = [
        "http://download.tensorflow.org/models/mobilenet_v1_2018_08_02/
mobilenet_v1_1.0_224_quant.tgz",
        "http://download.tensorflow.org/models/mobilenet_v1_2018_08_02/
mobilenet_v1_1.0_224_quant.tgz",
    ],
)
```

這段程式的意義是從 urls 指定的路徑下載並解壓縮檔，並用 flite_
mobilenet_ quant.BUILD 去建置新的目標。下面是 flite_mobilenet_quant.
BUILD 的程式，它做的就是對於解壓縮後的檔案，忽略 build 檔案，並
把其他檔案曝露出來，以便其他建置目標可以參照這些檔案。這段程式
裡有兩個相同的 URL，希望下個版本能夠改正。

```
exports_files(
    glob(
        ["**/*"],
        exclude = [
            "BUILD",
        ],
```

```
    ),
)
```

如 果 下 載 download.tensorflow.org/models/mobilenet_v1_2018_08_02/ 下
的 mobilenet_ v1_1.0_224_quant.tgz 並解壓，我們可以獲得

```
-rw-r----- 1 17020468 Aug  218:38 mobilenet_v1_1.0_224_quant.ckpt. data-
00000-of-00001
-rw-r----- 1    14644 Aug  218:38 mobilenet_v1_1.0_224_quant.ckpt.index
-rw-r----- 1  5143394 Aug  218:38 mobilenet_v1_1.0_224_quant.ckpt.meta
-rw-r----- 1   885850 Aug  218:38 mobilenet_v1_1.0_224_quant_eval.pbtxt
-rw-r----- 1 17173742 Aug  218:38 mobilenet_v1_1.0_224_quant_frozen.pb
-rw-r----- 1       89 Aug  218:38 mobilenet_v1_1.0_224_quant_info.txt
-rw-r----- 1  4276352 Aug  218:39 mobilenet_v1_1.0_224_quant.tflite
-rw-r----- 1 35069912 Aug  219:01 mobilenet_v1_1.0_224_quant.tgz
```

其中 mobilenet_v1_1.0_224_quant.tflite 就是應用中要使用的模型檔案。

TensorFlow Lite Android 應用的程式和 TensorMobile 的程式在應用層面
十分類似，例如：

- 讀取攝影機的影像。
- 讀取音訊資料。
- 產生執行緒驅動模型。
- 儲存模型。
- 讀取模型。
- 使用者介面。

兩者的不同點主要是模型的不同和呼叫模型的方法不同，以及有些模型
的輸入和輸出不同。我們在第 7 章說明了應用層面的程式，以及呼叫模
型的方法，這裡就不重複了。下面，我們主要講一下 TensorFlow Lite 的
幾個應用和它們的視覺化模型視圖。

先做些準備工作：

```
$ bazel build tensorflow/lite/tools:visualize
```

然後就可以使用建置後獲得的 visualize 指令稿工具，把 TensorFlow Lite 模型轉換輸出為 HTML 檔案，標頭檔案很多資訊，例如模型視圖、Ops 等，我們下面會做詳細説明。

7.3.1 聲音識別

1 模型

模 型 可 以 從 https://mirror.bazel.build/storage.googleapis.com/download. tensorflow.org/ models/tflite/conv_actions_tflite.zip 下載，解壓後可以看到以下檔案：

```
-rw-rw-r-- 1 3771180 Mar  8  2018 conv_actions_frozen.tflite
-r--r--r-- 1      60 Mar  8  2018 conv_actions_labels.txt
-rw-r----- 1 3494186 Apr  2  2018 conv_actions_tflite.zip
```

使用上面提到的 visualize 指令稿，執行下面的指令：

```
tensorflow/lite/tools/visualize conv_actions_frozen.tflite result.html
```

獲得 result.html，然後在瀏覽器裡載入這個檔案，可以獲得如圖 7-5 所示的聲音識別模型視圖。

輸出的 HTML 檔案裡有很多資訊，模型的概要資訊（模型的名稱、這個模型是如何獲得的）如下：

```
filename conv_actions_frozen.tflite version 3 description TOCO Converted.
```

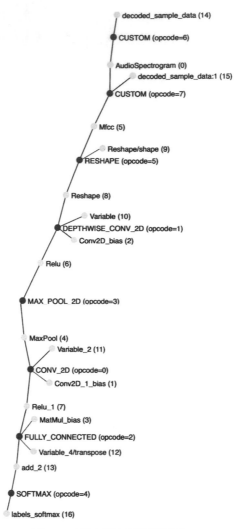

圖 7-5 聲音識別模型視圖

模型的輸入輸出的資訊如下：

```
Subgraph 0
Inputs      Outputs
Inputs      outputs
[14, 15     [16]
```

從上面這些資訊可以看到，輸入是 14 和 15 張量，輸出是 16 張量，這些數字對應的資訊如表 7-1 所示。

表 7-1 Tensor

index	name	type	shape	buffer	quantization
0	AudioSpectrogram	FLOAT32	[]	12	{details_type: 0, quantized_dimension: 0}
1	Conv2D_1_bias	FLOAT32	[64]	2	{details_type: 0, quantized_dimension: 0}
2	Conv2D_bias	FLOAT32	[64]	8	{details_type: 0, quantized_dimension: 0}
3	MatMul_bias	FLOAT32	[12]	9	{details_type: 0, quantized_dimension: 0}
4	MaxPool	FLOAT32	[]	17	{details_type: 0, quantized_dimension: 0}
5	Mfcc	FLOAT32	[]	13	{details_type: 0, quantized_dimension: 0}
6	Relu	FLOAT32	[]	14	{details_type: 0, quantized_dimension: 0}
7	Relu_1	FLOAT32	[]	5	{details_type: 0, quantized_dimension: 0}
8	Reshape	FLOAT32	[]	15	{details_type: 0, quantized_dimension: 0}
9	Reshape/shape	INT32	[4]	16	{details_type: 0, quantized_dimension: 0}
10	Variable	FLOAT32	[1, 20, 8, 64]	7	{details_type: 0, quantized_dimension: 0}
11	Variable_2	FLOAT32	[64, 10, 4, 64]	6	{details_type: 0, quantized_dimension: 0}
12	Variable_4/transpose	FLOAT32	[12, 64000]	1	{details_type: 0, quantized_dimension: 0}
13	add_2	FLOAT32	[]	4	{details_type: 0, quantized_dimension: 0}
14	decoded_sample_data	FLOAT32	[16000, 1]	11	{details_type: 0, quantized_dimension: 0, min: [0.0], max: [255.0]}

index	name	type	shape	buffer	quantization
15	decoded_sample_data:1	INT32	[1]	10	{details_type: 0, quantized_dimension: 0, min: [0.0], max: [255.0]}
16	labels_softmax	FLOAT32	[]	3	{details_type: 0, quantized_dimension: 0}

表 7-1 列出了所有的張量 Tensor 和對應的索引。如表 7-2 所示列出了所有的 Ops 和它們的輸入輸出的張量。

表 7-2 Ops

index	inputs	outputs	builtin_options	opcode_index
0	[14]	[0]	None	CUSTOM (opcode=6)
1	[0, 15]	[5]	None	CUSTOM (opcode=7)
2	[5, 9]	[8]	{new_shape: [-1, 99, 40, 1]}	RESHAPE (opcode=5)
3	[8, 10, 2]	[6]	{dilation_w_factor: 1, stride_h: 1, stride_w: 1, fused_activation_function: 'REL', depth_multiplier: 64, padding: 'SAME', dilation_h_factor: 1}	DEPTHWISE_CONV_2D (opcode=1)
4	[6]	[4]	{filter_height: 2, filter_width: 2, stride_h: 2, stride_w: 2, fused_activation_function: 'NONE', padding: 'SAME'}	MAX_POOL_2D (opcode=3)
5	[4, 11, 1]	[7]	{dilation_w_factor: 1, stride_h: 1, stride_w: 1, fused_activation_function: 'REL', padding: 'SAME', dilation_h_factor: 1}	CONV_2D (opcode=0)
6	[7, 12, 3]	[13]	{weights_format: 'DEFAULT', fused_activation_function: 'NONE'}	FULLY_CONNECTED (opcode=2)
7	[13]	[16]	{beta: 1.0}	SOFTMAX (opcode=4)

如表 7-3 所示列出了每個張量所需儲存的大小。

表 7-3 Buffer

index	data	index	data
0	--	9	48 bytes
1	3072000 bytes	10	--
2	256 bytes	11	--
3	--	12	--
4	--	13	--
5	--	14	--
6	655360 bytes	15	--
7	40960 bytes	16	16 bytes
8	256 bytes	17	--

演算子的資訊如表 7-4 所示。請注意 6 和 7 是訂製的演算子。

表 7-4 Operator Code

index	builtin_code	custom_code
0	CONV_2D	None
1	DEPTHWISE_CONV_2D	None
2	FULLY_CONNECTED	None
3	MAX_POOL_2D	None
4	SOFTMAX	None
5	RESHAPE	None
6	CUSTOM	AudioSpectrogram
7	CUSTOM	Mfcc

7.3.2 影像識別

前面的應用使用了 SSD Mobilenet 模型,該模型非常大,不能在書裡全部展示,取出其中的一部分如圖 7-6 所示,供讀者參考。

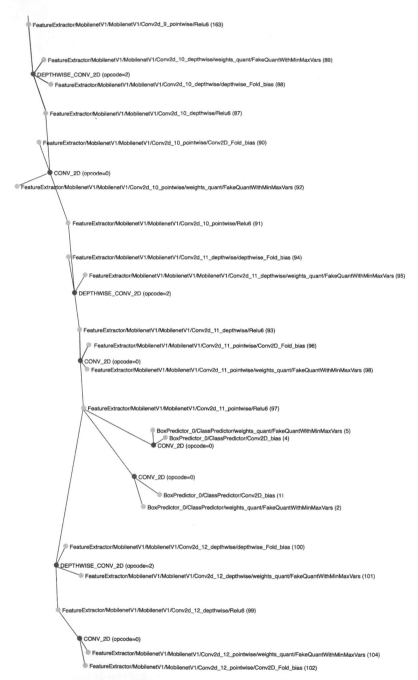

圖 7-6 影像識別模型視圖

由於模型資料非常大,詳細內容就不寫進本書了,有興趣的讀者可以嘗試執行一下 Visualize 工具,自行檢視視圖。

7.4 TensorFlow Lite 使用 GPU

由於行動裝置的處理能力有限,在行動裝置上執行以計算密集型為主的機器學習模型的推理,對運算資源的要求很高。雖然轉為定點模型是加速的一種途徑,但是運算效能還是不高。TensorFlow Lite 近期發佈了對 GPU 的支援,使用 GPU 不會對程式產生額外的複雜性和潛在的量化精度損失。

TensorFlow Lite 對 GPU 的支援是透過 OpenGL ES 3.1 計算著色器(Compute Shaders)來實現的。讀者可以下載並嘗試使用 GPU 的預先編譯二進位預覽版,TensorFlow Lite 官方許諾接下來會發佈一個完整的開放原始碼版本。目前很多 TensorFlow Lite 的應用仍使用 CPU 進行運算,例如人臉輪廓檢測,未來我們會利用新的 GPU 後端,預計可將 Pixel3 和三星 S9 的推理速度提升 4~6 倍。

7.4.1 GPU 與 CPU 效能比較

Google 公佈了一些他們的測試結果,這些結果是在 Google 的產品中進行幾個月的測試後獲得的。對於 Pixel3 的肖像模式,TensorFlow Lite GPU 讓前景 - 背景分割模型的速度加強了 4 倍以上,新的深度預估模型的推理速度加強了 10 倍以上。在 YouTube 上的 YouTube Stories 和 Playground Stickers 中,即時視訊分割模型在各種手機上的速度加強了 5~10 倍。對於各種各樣的深度神經網路模型,新的 GPU 後端通常比浮點 CPU 實現速度快 2~7 倍。

7.4.2 開發 GPU 代理（Delegate）

GPU 旨在為大規模可平行化的計算工作負載提供高傳輸量。因此，GPU 非常適用於深度神經網路，深度神經網路由大量運算子組成，每個運算子處理一些輸入張量，這些輸入張量可以容易地分成較小的工作單元並且並存執行，這樣做通常可以加強計算的延遲。在最佳情況下，GPU 上的推理運算現在足夠快，可用於以前不可能實現的即時應用程式。

與 CPU 不同，GPU 使用 16 位元或 32 位元浮點數進行計算，並且不需要量化以獲得最佳效能。GPU 推理的另一個好處是它的功效。GPU 以非常有效和最佳化的方式執行計算，因此與在 CPU 上執行相同工作相比，它們消耗更少的功率並產生更少的熱量。

1 示範應用程式

現在我們嘗試在 Android 上開發和整合支援 GPU 的 TensorFlow Lite 的預覽版本。

這裡我們使用 Gradle 來建置，有興趣的讀者可以嘗試使用 Bazel 的 BUILD 來整合建置。實作方式程式如下：

```
dependencies {
    implementation 'org.tensorflow:tensorflow-lite:0.0.0-gpu-experimental'
}
```

接下來，在應用裡增加 GPU 的支援程式，程式很簡單，就是產生一個 GpuDelegate：

```
import org.tensorflow.lite.Interpreter;
import org.tensorflow.lite.experimental.GpuDelegate;

delegate = new GpuDelegate();
```

```
Interpreter.Options options = (new Interpreter.Options()).addDelegate
(delegate);
Interpreter interpreter = new Interpreter(model, options);

// 執行推理
while (true) {
  writeToInput(input);
  interpreter.run(input, output);
  readFromOutput(output);
}

// 清理空間
delegate.close();
```

2 GPU 支援的模型和運算元

下面是 GPU 現在支援的模型的清單：

- MobileNet v1（224x224）影像分類器。它是為行動裝置和嵌入式裝置而設計的影像分類模型。

- DeepLab Segmentation（257x257）影像分割模型，將語義標籤（例如狗、貓、汽車）分配給輸入影像中的每個像素。

- MobileNet SSD Object Detection 影像分類模型，用於檢測帶有邊界框的多個物體。

- PoseNet for Pose Estimation 視覺模型，用於估計影像或視訊。

GPU 上的 TFLite 支援 16 位元和 32 位元浮點精度的以下操作：

- ADD
- AVERAGE_POOL_2D
- CONCATENATION

- CONV_2D
- DEPTHWISE_CONV_2D
- FULLY_CONNECTED
- LOGISTIC
- LSTM v2 (Basic LSTM only)
- MAX_POOL_2D
- MUL
- PAD
- PRELU
- RELU
- RELU6
- RESHAPE
- RESIZE_BILINEAR
- SOFTMAX
- STRIDED_SLICE
- SUB
- TRANSPOSE_CONV

❸ 不支援的模型和操作

如果 GPU 的代理不支援某些操作，則 TFLite 將僅在 GPU 上執行圖形中可支援的 Ops，而在 CPU 上執行其餘部分。由於 CPU/GPU 同步的成本高，像這樣的拆分執行模式通常會導致效能比單獨在 CPU 上執行整個網路時慢。在這種情況下，程式會產生以下警告：

```
WARNING: op code #42 cannot be handled by this delegate.
```

程式沒有為此失敗提供回呼函數，因為這不是真正的執行時錯誤，而是開發人員在嘗試讓網路在 GPU 代理上執行時期可以觀察到的內容。

▓4▓ 實現原理

深度神經網路會按順序執行數百個操作，使它們非常適合 GPU，GPU 的設計考慮了針對高傳輸量的平行工作負載。

Interpreter::ModifyGraphWithDelegate() 在 C++ 中呼叫時初始化 GPU，或透過 Interpreter.Options 間接呼叫 Interpreter 的建置函數來初始化 GPU。下面是用 C++ 實現的參考程式：

```cpp
auto model = FlatBufferModel::BuildFromFile(model_path);
tflite::ops::builtin::BuiltinOpResolver op_resolver;
std::unique_ptr<Interpreter> interpreter;
InterpreterBuilder(*model, op_resolver)(&interpreter);

auto* delegate = NewGpuDelegate(nullptr);  // default config
QCHECK_EQ(interpreter->ModifyGraphWithDelegate(delegate), kTfLiteOk);

WriteToInputTensor(interpreter->typed_input_tensor<float>(0));
QCHECK_EQ(interpreter->Invoke(), kTfLiteOk);
ReadFromOutputTensor(interpreter->typed_output_tensor<float>(0));

DeleteGpuDelegate(delegate);
```

在初始化階段，以架構接收為基礎的執行計畫（Execution Plan）建置輸入神經網路的表示。使用這種新的表示，可以實現一些有用的轉換，這些轉換包含以下內容：

- 剔除不需要的 Ops。

- 將 Ops 取代為效能更好的相等 Ops。

- 合併 Ops，以減少最後產生的著色程式的數量。

然後基於此最佳化圖（Optimized Graph），產生並編譯計算著色器。在 Android 上現在使用 OpenGL ES 3.1 計算著色器，在建立這些計算著色器時，程式還採用了各種結構的最佳化，例如：

■ 應用某些操作的特化而非它們的通用實現。

■ 釋放暫存器壓力。

■ 選擇最佳工作群組大小。

■ 安全地調整精度。

■ 重新排序顯性數學操作。

在這些最佳化結束時，著色器程式被編譯，可能需要幾毫秒到半秒，就像手機遊戲一樣。一旦著色程式編譯完成，新的 GPU 推理引擎就可以開始工作了。

在對每個輸入進行推理運算時：

■ 如有必要，輸入將移至 GPU。輸入張量（如果尚未儲存為 GPU 記憶體）可由架構透過建立 GL 緩衝區或 MTLBuffers 進行 GPU 存取，同時還可能複製資料。由於 GPU 在 4 通道資料結構中效率最高，因此通道大小不等於 4 的張量將重新調整為更加適合 GPU 的架構。

■ 執行著色器程式。將上述著色器程式插入指令緩衝區佇列中，GPU 將這些程式輸出。在此步驟中，我們還為中間張量管理 GPU 記憶體，以盡可能減少後端的記憶體佔用。

■ 必要時將輸出移動到 CPU。一旦深度神經網路處理完成，架構將結果從 GPU 記憶體複製到 CPU 記憶體，除非網路的輸出可以直接在螢幕上呈現，不需要這樣的傳輸。

7.5 訓練模型

TensorFlow Lite 主要是針對推理（Inference）的，但是，隨著硬體裝置的發展和技術的進步，未來在行動裝置上進行模型訓練並商業化，應該也不是不可能的事。我們在這裡介紹一個簡單實現的程式。

7.5.1 模擬器

首先，準備模擬器（Emulator）。在安裝 Android SDK 時，請同時安裝 Intel x86 Atom System 映射。

然後，讓我們列出可用的 Android 模擬器，AVD 是 Android Virtual Devices 的簡稱。

```
$ /opt/Android/sdk/tools/android list avd
************************************************************************
The "android" command is deprecated.
For manual SDK, AVD, and project management, please use Android Studio.
For command-line tools, use tools/bin/sdkmanager and tools/bin/avdmanager
************************************************************************
Running /opt/Android/sdk/tools/bin/avdmanager list avd
```

或，執行下面的指令：

```
$ /opt/Android/sdk/tools/bin/avdmanager list avd

Available Android Virtual Devices:
    Name: Pixel_API_26
  Device: pixel (Google)
    Path: .android/avd/Pixel_API_26.avd
  Target: Google APIs (Google Inc.)
          Based on: Android 8.0 (Oreo) Tag/ABI: google_apis/x86
```

```
    Skin: pixel
  Sdcard: 100M
```

現在，我們可以選擇一個模擬器並執行它，例如執行下面的指令：

```
/opt/Android/sdk/tools/emulator -avd Pixel_API_26
```

7.5.2 建置執行檔案

在 ensorflow/cc/build 中，增加以下目標：

```
cc_binary(
    name = "android_tutorials_example_trainer",
    srcs = ["tutorials/example_trainer.cc"],
    copts = tf_copts(),
    linkopts = select({
        "//tensorflow:android": [
            "-lm",
            "-fPIE",
            "-pie",
            "-llog",
            "-latomic", # x86 requires
            "-landroid",
            "-Wl,-z,defs",
            "-Wl,--no-undefined",
            "-s",
        ],
        "//conditions:default": [
            "-lm",
            "-lpthread",
            "-lrt",
        ],
    }),
    deps = [
```

```
        "while_loop",
        "//tensorflow/core:android_tensorflow_lib",
    ],
)
```

現在，我們可以透過下面的指令來建置執行檔案：

```
$ bazel build -c opt --cxxopt='--std=c++11' --crosstool_top=//external:
android/crosstool --host_crosstool_top=@bazel_tools//tools/cpp:toolchain
--config=android --cpu=x86 //tensorflow/cc:android_tutorials_example_trainer
```

獲得的執行檔案如下：

```
bazel-bin/tensorflow/cc/android_tutorials_example_trainer
```

為使我們有權將可執行檔複製到行動裝置，執行以下指令：

```
adb root
```

然後，複製檔案並執行：

```
$ adb push bazel-bin/tensorflow/cc/android_tutorials_example_trainer/data/
local/tmp/
```

最後，執行訓練，執行指令及結果如下：

```
$ adb shell /data/local/tmp/android_tutorials_example_trainer
native : cpu_feature_guard.cc:35 The TensorFlow library was compiled to use SSE
instructions, but these aren't available on your machine.
native : cpu_feature_guard.cc:35 The TensorFlow library was compiled to use SSE2
instructions, but these aren't available on your machine.
native : cpu_feature_guard.cc:35 The TensorFlow library was compiled to use SSE3
instructions, but these aren't available on your machine.
can't determine number of CPU cores: assuming 4
can't determine number of CPU cores: assuming 4
```

```
000000/000001 lambda = 3.056837 x = [0.9692450.246098] y = [3.399931 -0.969245]
000000/000003 lambda = 1.716893 x = [0.6391000.769124] y = [3.455547 -0.639100]
000000/000003 lambda = 2.721944 x = [0.983324 -0.181865] y = [2.586241 -0.983324]
000000/000003 lambda = 2.288898 x = [0.934717 -0.355392] y = [2.093368 -0.934717]
000000/000003 lambda = 2.129015 x = [0.913109 -0.407716] y = [1.923896 -0.913109]
000000/000003 lambda = 2.061103 x = [0.903412 -0.428773] y = [1.852692 -0.903412]
000000/000003 lambda = 2.029754 x = [0.898833 -0.438291] y = [1.819919 -0.898833]
000000/000003 lambda = 2.014684 x = [0.896609 -0.442823] y = [1.804181 -0.896609]
000000/000003 lambda = 2.007294 x = [0.895513 -0.445036] y = [1.796467 -0.895513]
000000/000003 lambda = 2.003635 x = [0.894969 -0.446129] y = [1.792648 -0.894969]
000000/000003 lambda = 2.001815 x = [0.894698 -0.446672] y = [1.790748 -0.894698]
000000/000003 lambda = 2.000907 x = [0.894562 -0.446943] y = [1.789801 -0.894562]
000000/000003 lambda = 2.000453 x = [0.894495 -0.447078] y = [1.789327 -0.894495]
000000/000003 lambda = 2.000227 x = [0.894461 -0.447146] y = [1.789091 -0.894461]
... ... ...
000000/000005 lambda = 2.000000 x = [0.894427 -0.447214] y = [1.788854 -0.894427]
000000/000005 lambda = 2.000000 x = [0.894427 -0.447214] y = [1.788854 -0.894427]
000000/000005 lambda = 2.000000 x = [0.894427 -0.447214] y = [1.788854 -0.894427]
```

行動端的機器學習開發

本章前半部分簡介 TensorFlow 對其他行動平台和嵌入式平台的支援，後半部分介紹行動端 TensorFlow 開發的實戰經驗。

8.1 其他裝置的支援

下面介紹 TensorFlow 對 iOS 和樹莓派的支援。iOS 在行動端佔了很大百分比，是每個開發人員都會關注的平台。樹莓派作為 IoT 和小型裝置的代表，在其上面進行行動開發有很重要的示範意義。

8.1.1 在 iOS 上執行 TensorFlow 的應用

我們已經在 iOS 上看到了很多使用 TensorFlow 的優秀的應用程式，iOS 大概佔有 20% 的市佔率，所以支援這個平台對 TensorFlow Lite 很重要。

如果讀者需要進行開發，TensorFlow 提供了完整的從原始程式碼建置到開發應用的方法。

1 使用 CocoaPods

在 iOS 上使用 TensorFlow 最簡單的方法是使用 CocoaPods 套件管理系統。讀者可以從 cocoapods.org 上下載 TensorFlow 架構套件，然後只需執行 "TensorFlow-experimental"，將其作為依賴增加到你的應用程式的 Xcode 專案中。這樣就安裝了一個通用的二進位架構，這使得開發者很容易入門，但缺點是開發者難以訂製軟體套件，這對於縮小應用的二進位大小非常重要。

開發者可以執行下面的指令稿，完成安裝。

```
tensorflow/contrib/makefile/build_all_ios.sh
```

這個過程大約需要 20 分鐘。

2 使用 Makefile Unix

make 程式可能是最古老的實用建置工具之一，用 make 直接建置和操作檔案，這種較底層的操作方式為一些棘手的開發情況提供了很多的靈活性，例如在交換編譯或在舊的及有限的資源系統上建置時，TensorFlow 在 tensorflow/contrib/makefile 中提供了一個針對行動和嵌入式平台的 Makefile。這是建置 iOS 的主要方式，針對 Linux、Android 和樹莓派也是有用的。下面介紹一下使用這種方法進行手動建置的過程。

首先，執行下面的指令稿，下載依賴的軟體套件：

```
tensorflow/contrib/makefile/download_dependencies.sh
```

然後，執行下面的指令稿為 iOS 編譯 ProtoBuf：

```
tensorflow/contrib/makefile/compile_ios_protobuf.sh
```

接著，執行指定 iOS 作為目標的 Makefile，以及 iOS 相對應的晶片架構：

```
make -f tensorflow/contrib/makefile/Makefile \
    TARGET=IOS \
    IOS_ARCH=ARM64
```

執行結束後，會產生 tensorflow/contrib/makefile/gen/lib/libtensorflow-core.a 的通用軟體函數庫，開發人員可以使用這個函數庫連結任何 Xcode 專案。

3 編譯最佳化

compile_ios_tensorflow.sh 指令稿可以接受可選的命令列參數。第一個參數是 C++ 最佳化參數，預設為偵錯模式。開發者可以設定更高的最佳化參數，如下所示：

```
compile_ios_tensorflow.sh "-Os"
```

有關最佳化參數的其他選項，請參見最佳化等級的文件。

iOS 範例有三個示範應用程式，全部定義在 tensorflow/contrib/ios_examples 中的 Xcode 專案中。

4 簡單示範

示範如何用最少的程式載入和執行 TensorFlow 模型的最簡單範例。它由一個單一的視圖和一個按鈕組成，當按下按鈕時，會執行模型載入和推理。

5 相機示範

類似 Android 的 TensorFlow 影像分類示範。它載入了 Inception v3，並輸出其最佳影像標籤，也可以即時檢視相機視圖中的內容。與 Android 版本一樣，開發者可以使用 TensorFlow for Poets（https://codelabs.developers.google.com/codelabs/tensorflow-for-poets）

來訓練開發者自訂的模型，並將其放入此範例中，只需更改少量程式即可。

6 效能測試示範

與簡單示範很接近，但它重複執行模型，並將統計資料輸出給基準測試（Benchmark）工具。要建置這些示範，首先要確保你已經能夠成功地為 iOS 編譯主 TensorFlow 函數庫。開發者還需要下載需要的模型檔案，請執行下面的程式：

```
mkdir -p ~/graphs
curl -o ~/graphs/inception5h.zip \
    https://storage.googleapis.com/download.tensorflow.org/models/
inception5h.zip
    unzip ~/graphs/inception5h.zip -d ~/graphs/inception5h
cp ~/graphs/inception5h/* \
    tensorflow/examples/ios/benchmark/data/
cp ~/graphs/inception5h/* \
    tensorflow/examples/ios/camera/data/
cp ~/graphs/inception5h/* \
    tensorflow/examples/ios/simple/data
```

開發者應該能夠為每個單獨的示範載入 Xcode 專案，建置它並在裝置上執行。相機示範需要一個相機，所以它不會在模擬器上執行，但影像繪製和聲音識別應用應該可以在模擬器上執行。開發者也可以直接從 iOS

應用程式直接使用 C++，程式可以直接呼叫 TensorFlow 架構的 API。

8.1.2 在樹莓派上執行 TensorFlow

TensorFlow 團隊正在努力為樹莓派提供一個官方的整合套件的安裝路徑，以便使用預先建置的二進位檔案在樹莓派上執行該架構。在寫這篇文章的時候它還不能用（檢視 https://www.tensorflow.org/install 了解最新的細節），在這裡筆者將介紹如何從原始程式碼建置它。

以樹莓派為基礎的建置與普通的 Linux 系統類似。首先，下載並安裝依賴的軟體套件，建置 ProtoBuf：

```
tensorflow/contrib/makefile/download_dependencies.sh
sudo  apt-get install -y \
autoconf automake libtool gcc-4.8 g++-4.8
cd tensorflow/contrib/makefile/downloads/protobuf/
./autogen.sh
./configure
make
sudo make install
sudo ldconfig # refresh shared library cache
cd ../../../../..
```

然後，使用 make 指令來建立函數庫和應用實例：

```
make -f tensorflow/contrib/makefile/Makefile HOST_OS=PI \
     TARGET=PI OPTFLAGS="-Os -mfpu=neon-vfpv4 \
      -funsafe-math-optimizations -ftree-vectorize" CXX=g++-4.8
```

下面介紹樹莓派的應用實例。

樹莓派是各種嵌入式應用原型的絕佳平台。在 tensorflow/contrib/pi_examples 中有兩個不同的實例：

1 影像標籤

影像標籤是標準的 tensorflow/examples/label_image 示範的移植示範，它嘗試利用 Inception v3 Imagenet 的模型來標記影像。與其他平台一樣，開發者可以輕鬆地使用從 TensorFlow for Poets 衍生的訂製訓練版本來取代此模型。

2 相機

這個實例使用樹莓派的攝影機 API 來讀取一個即時視訊，在其上執行影像標籤，並將標籤輸出到主控台。為了讓示範更有趣，此應用可以讓開發者將結果提供給 tflite 文字語音工具，以便讓樹莓派能說出所看到的內容。要建置這些範例，請確保你已經執行了前面所示的樹莓派建置過程，然後執行以下指令：

```
$ makefile -f tensorflow/contrib/pi_examples/camera
```

或執行以下指令：

```
$ makefile -f tensorflow/contrib/pi_examples/simple
```

執行完指令將產生一個根資料夾 gen/bin，執行其中的可執行檔即可取得模型檔案，範例程式如下：

```
curl https://storage.googleapis.com/download.tensorflow.org/ \
        models/inception_dec_2015_stripped.zip \
        -o /tmp/inception_dec_2015_stripped.zip
    unzip /tmp/inception_dec_2015_stripped.zip \
        -d tensorflow/contrib/pi_examples/label_image/data/
```

8.2 設計和最佳化模型

以上我們學習和研究了 TensorFlow Mobile 和 TensorFlow Lite，並大致了解 iOS 和樹莓派的開發，下面我們歸納一下如何在行動端和嵌入式裝置上進行機器學習的開發，以及開發中需要注意的地方。當開發者嘗試在行動裝置或嵌入式裝置上發佈應用時，有一些特殊問題需要處理，例如怎樣最佳化延遲，最佳化記憶體（RAM）的使用，最佳化模型檔案大小和執行檔案二進位大小，開發者在開發模型時也需要考慮這些問題。

8.2.1 模型大小

模型需要儲存在裝置上的某個地方，大型的神經網路可能需要數百百萬位元組的儲存空間。即使有足夠的本機存放區空間，當使用者需要從市集下載非常大的應用套裝程式時，就會佔據很多儲存空間，因此開發者需要規劃模型的規模。

在 使 用 freeze_graph 和 strip_unused_nodes 之 後， 開 發 者 可 以 檢視 GraphDef 檔案在磁碟上的大小，因為它應該只包含推理相關的節點。要仔細檢查模型是否符合預期，開發者可以透過 summarize_graph 指令檔來檢視常數中有多少個參數：

```
bazel build tensorflow/tools/graph_transforms:summarize_graph \
  && bazel-bin/tensorflow/tools/graph_transforms/summarize_graph \
  --in_graph=/tmp/tensorflow_inception_graph.pb
```

程式執行結果如下：

```
No inputs spotted.
Found 1 possible outputs: (name=softmax, op=Softmax)
Found 23885411 (23.89M) const parameters, 0 (0) variable
```

```
 parameters, and 99 control_edges
Op types used: 489 Const, 99 CheckNumerics, 99 Identity,
 94 BatchNormWithGlobalNormalization, 94 Conv2D, 94 Relu,
 11 Concat, 9 AvgPool, 5 MaxPool, 1 Sub, 1 Softmax,
 1 ResizeBilinear, 1 Reshape, 1 Mul, 1 MatMul, 1 ExpandDims,
 1 DecodeJpeg, 1 Cast, 1 BiasAdd
```

注意，結果中常數參數的個數是 23 885 411，一般每個常數參數是以 32 位元浮點數儲存的，我們用參賽個數乘以 4，獲得的數值應該和模型檔案的大小比較接近。如果我們把 32 位數值改成 8 位元數值，那麼模型的損失會非常小，可以省下很多的儲存空間。如果模型過大，開發者可以採用這種方法減少模型的大小。實作方式程式如下：

```
bazel build tensorflow/tools/graph_transforms:transform_graph && \
bazel-bin/tensorflow/tools/graph_transforms/transform_graph \
--in_graph=/tmp/tensorflow_inception_optimized.pb \
--out_graph=/tmp/tensorflow_inception_quantized.pb \
--inputs='Mul:0' --outputs='softmax:0' --transforms='quantize_weights'
```

註：參數中的 "--transforms='quantize_weights' " 表示使用定點化的方法。

另外一種模型壓縮方法是 round_weights。這種方法不能壓縮模型本身的大小，但是可以使模型在壓縮後的檔案更小。這種方法將加權參數儲存為浮點數，但將其捨入為設定的步數值。這表示儲存模型中有更多的重複位元組模式，所以壓縮常常會大幅降低檔案的大小。

在很多情況下，最後的檔案大小可能會非常接近 8 位元模型。這樣做的優點是，架構不必分配一個臨時緩衝區來解壓縮參數，就像我們在使用 quantize_weights 時一樣。這樣可以適當節省執行的延遲。

我們還可以使用記憶體對映，以此減少載入模型的時間，這在本書的第 7 章中講過。

8.2.2 執行速度

大多數模型開發和部署的最高優先順序之一就是，如何快速執行推理以提供良好的使用者體驗。這個過程的第一步是，檢視執行圖所需的浮點運算的總數。開發者可以透過使用 benchmark_model 指令稿工具來得到一個非常粗略的估算，程式如下：

```
bazel build -c opt tensorflow/tools/benchmark:benchmark_model \
 && bazel-bin/tensorflow/tools/benchmark/benchmark_model \
 --graph=/tmp/inception_graph.pb --input_layer="Mul:0" \
 --input_layer_shape="1,299,299,3" --input_layer_type="float" \
 --output_layer="softmax:0" \
 --show_run_order=false --show_time=false \
 --show_memory=false --show_summary=true --show_flops=true \
 --logtostderr
```

執行這個指令稿之後，就會顯示需要多少操作來執行 TensorFlow 模型圖。然後開發者就可以透過這些資訊來確定模型在目標裝置上執行的可行性。舉個實例，2016 年的高階手機每秒鐘可以處理 200 億個 FLOP，所以執行一個需要 100 億個 FLOP 的模型的最佳速度大約是 500 毫秒。在像樹莓派 3 那樣可以做大約 50 億個 FLOP 的裝置上，大概每兩秒鐘只能獲得一個推理結果。

有了這個估計之後就可以規劃在行動裝置上可以實現的功能，如果模型使用的操作太多，那麼我們就要儘量最佳化架構以減少這個運算的數量。另外，也可以選擇比較先進的模型，例如 SqueezeNet 和 MobileNet。也可以尋找替代模型，甚至可能更小的舊模型。舉例來說，Inception v1 只有約 700 萬個參數，執行需要 90 億個 FLOP。而 Inception v3 的 2400 萬個參數執行需要 30 億個 FLOP。

如何檢驗你的模型？

一旦開發者了解了行動裝置可能達到的最佳效能，我們就要看看應用能獲得的實際效能。我們推薦在比較獨立和隔絕的空間裡執行應用，不推薦和其他應用程式混合執行，因為這有助隔離 TensorFlow 對延遲的影響。TensorFlow 的基準工具（Benchmark）可以幫助開發者做到這一點。如果要檢驗 Inception v3 的效能，請執行以下程式：

```
bazel build -c opt tensorflow/tools/benchmark:benchmark_model \
 && bazel-bin/tensorflow/tools/benchmark/benchmark_model \
 --graph=/tmp/tensorflow_inception_graph.pb \
 --input_layer="Mul" --input_layer_shape="1,299,299,3" \
 --input_layer_type="float" --output_layer="softmax:0" \
 --show_run_order=false --show_time=false \
 --show_memory=false --show_summary=true \
 --show_flops=true --logtostderr
```

輸出結果如下：

```
========================= Top by Computation Time =========================
[node type] [start] [first] [avg ms] [ % ]  [cdf%]  [mem KB]   [Name]
Conv2D      22.859  14.212  13.700   4.972% 4.972%  3871.488 conv_4/Conv2D
Conv2D       8.116   8.964  11.315   4.106  9.078%  5531.904 conv_2/Conv2D
Conv2D      62.066  16.504   7.274   2.640% 11.717%  443.904 mixed_3/conv/
                                                                      Conv2D
Conv2D       2.530   6.226   4.939   1.792% 13.510% 2765.952 conv_1/Conv2D
Conv2D      55.585   4.605   4.665   1.693% 15.203%  313.600 mixed_2/
                                                             tower/conv_1/Conv2D
Conv2D     127.114   5.469   4.630   1.680% 16.883%   81.920 mixed_10/
                                                                  conv/Conv2D
Conv2D      47.391   6.994   4.588   1.665% 18.548%  313.600 mixed_1/
                                                             tower/conv_1/Conv2D
Conv2D      39.463   7.878   4.336   1.574% 20.122%  313.600 mixed/tower/
                                                                  conv_1/Conv2D
Conv2D     127.113   4.192   3.894   1.413% 21.535%  114.688 mixed_10/
                                                             tower_1/conv/Conv2D
```

```
Conv2D       70.188    5.205    3.626  1.316%  22.850%    221.952  mixed_4/conv/
Conv2D
========================= Summary by node type =========================
[Node type]   [count]   [avg ms]   [avg %]      [cdf%]        [mem KB]
Conv2D          94      244.899    88.952%     88.952%      35869.953
BiasAdd         95        9.664     3.510%     92.462%      35873.984
AvgPool          9        7.990     2.902%     95.364%       7493.504
Relu            94        5.727     2.080%     97.444%      35869.953
MaxPool          5        3.485     1.266%     98.710%       3358.848
Const          192        1.727     0.627%     99.337%          0.000
Concat          11        1.081     0.393%     99.730%       9892.096
MatMul           1        0.665     0.242%     99.971%          4.032
Softmax          1        0.040     0.015%     99.986%          4.032
<>               1        0.032     0.012%     99.997%          0.000
Reshape          1        0.007     0.003%    100.000%          0.000
Timings (microseconds): count=50 first=330849 curr=274803 min=232354
max=415352 avg=275563 std=44193
Memory (bytes): count=50 curr=128366400(all same)
514 nodes defined 504 nodes observed
```

上面是輸出的摘要,注意指令稿中 show_summary 的參數應設為真。輸出中第一個表格是按時間順序花費最多時間的節點列表。從左到右依次是:

- 節點類型,即什麼樣的操作。

- 操作的開始時間。

- 第一次運算的時間,以毫秒為單位。這是基準測試第一次執行需要多長時間,預設情況下會執行 20 次執行以獲得更可靠的統計資料。可以發現在第一次執行中進行較長的計算並且快取結果的操作是有用的。

- 所有執行的平均執行時間,以毫秒為單位。

- 執行一次所花時間的百分比。這對了解主要計算的發生位置非常有用。

- 表中這個操作和以前操作的累計總時間。這對於了解神經網路層間的工作分佈很方便,看看是否有少數節點在大部分時間都在被使用。

- 節點的名稱。

第二個表與第一個表類似,但不是按特定的運算節點來劃分時間,而是按照操作類型進行分組。這對想要了解從圖形中最佳化或消除哪些操作非常有用。該表只顯示前 10 個類型分類,從左到右依次是:

- 正在分析的節點的類型。

- 此類型的所有節點累積的平均時間,以毫秒為單位。

- 這種類型的操作佔總時間的百分比。

- 表中這個節點類型的累計時間較長,因此開發者可以了解工作負載的分佈情況。

- 此節點類型的輸出佔用了多少記憶體。

兩個表格都有作為列之間分隔符號的選項符號,因此可以輕鬆地將結果複製並貼上到試算表中。在尋找可以最佳化的點時,歸納節點類型可能是最有用的,因為它會告訴開發者哪些操作花費時間最多。在這種情況下,我們一般可以看到 Conv2D 的執行時間幾乎是執行時間的 90%。該跡象表明該圖是經過良好最佳化的,因為卷積和矩陣乘法計算預計佔了神經網路的計算工作量的大部分。

作為一個經驗法則,如果你看到其他操作佔用了很多時間,那是不太好的跡象。對於神經網路,不涉及大矩陣乘法的操作通常應該作為等級低的操作被處理。因此,如果開發者看到很多時間花費在這些操作上,則表明這個神經網路不是最佳結構,或實現這個神經網路的操作程式沒有被最佳化。遇到這種情況,開發者要儘量做效能缺陷修補程式。特別是，當你使用基準測試（Benchmark）工具看到了相似的問題時，更要注意。

上面的工具是執行在開發者的桌面電腦上的，但這些工具也可以執行在 Android 上，其實這是最適合行動開發的環境。以下是在 64 位元 ARM 裝置上執行它的範例程式：

```
bazel build -c opt --config=android_arm64 \
    tensorflow/tools/benchmark:benchmark_model
adb push bazel-bin/tensorflow/tools/benchmark/benchmark_model/data/local/ tmp
adb push /tmp/tensorflow_inception_graph.pb /data/local/tmp/
adb shell '/data/local/tmp/benchmark_model \
    --graph=/data/local/tmp/tensorflow_inception_graph.pb \
    --input_layer="Mul" --input_layer_shape="1,299,299,3" \
    --input_layer_type="float" --output_layer="softmax:0" \
    --show_run_order=false --show_time=false \
    --show_memory=false --show_summary=true'
```

對於 iOS 上可以執行的命令列工具，現在還沒有很好的支援，所以在 tensorflow/contrib/ ios_examples/benchmark 中有一個單獨的實例，它將應用程式中的相同功能包裝。這會將統計資訊輸出到裝置螢幕並進行記錄檔偵錯。

8.2.3 視覺化模型

加速開發者的程式最有效的方法是改變你的模型，它可以減少模型的計算量。要做到這一點，開發者需要了解模型正在做什麼，並將其作為視覺化的第一步。

為了高度概括模型的執行圖，建議開發者使用 Tensor Board 應用程式。它應該能夠載入開發者的 GraphDef 檔案，雖然有時可能會遇到較大的執

行圖。為了取得更細粒度的視圖,我們可以將圖形轉為 Graphviz 的 DOT
檔案格式。下面是相關的指令稿程式:

```
bazel build tensorflow/tools/quantization:graph_to_dot
bazel-bin/tensorflow/tools/quantization/graph_to_dot \
    --graph=/tmp/tensorflow_inception_graph.pb \
    --dot_output=/tmp/tensorflow_inception_graph.dot
```

如果讀者使用的是類 UNIX 環境,可以安裝 xdot 程式。執行 xdot 可能
需要花費幾秒鐘或更長的時間來載入非常大的執行圖。在預設情況下,
指令稿只顯示操作類型,因為名稱可能很長,螢幕輸出很難看。當然,
如果需要,也可以修改 Python 指令稿以顯示更多資訊。

8.2.4 執行緒

TensorFlow 的桌上版本具有複雜的執行緒模型,如果可以,將嘗試平行
執行多個操作,術語叫作「內部操作平行」。可以透過在階段(Session)
選項中指定 "interop_threads" 來設定操作間的平行性。

在行動裝置上,系統內部操作平行處理數(一次執行多少個運算元)預
設設定為 1,以便操作始終按順序連續執行。這種設定是合理的,因為
行動處理器通常具有很少的核心數和一個小的快取,所以執行多個存取
不相交部分記憶體的操作通常不會提高性能。操作內平行是非常有用
的,例如對於卷積運算,在多執行緒執行的情況下,可以為每個執行緒
分配一塊小的記憶體。

在行動裝置上,預設情況下,一個作業系統使用的執行緒數將被設定為
核心數量,或在核心數量無法確定的情況下,該數量將只有 4 個。讀者
可以透過使用階段選項顯性地設定它來覆蓋預設的執行緒數。如果你的

應用程式有自己的執行緒池，那麼它們就不會相互干擾，所以減少執行緒預設值是個好主意。

8.2.5 二進位檔案大小

行動裝置和桌面伺服器開發之間最大的區別是二進位檔案大小。在桌上型電腦上，數百百萬位元組的大型可執行檔並不罕見，但對行動和嵌入式應用程式來說，保持二進位檔案盡可能小是非常重要的，這樣使用者的下載將變得非常簡單和快速。

TensorFlow 的應用預設只包含 TensorFlowOps 的子集，但是這樣仍然導致最後可執行檔非常大。為了減少檔案的大小，開發者可以設定自動分析模型來設定函數庫，只包含應用實際需要的 Ops 的實現。要使用這種方法，請按照下列步驟操作：

（1）在你的模型上執行 tools/print_required_ops/ 目錄下的 print_selective_registration_header.py 指令稿，產生一個只啟用它所使用的操作的標頭檔。

（2）將 ops_to_register.h 檔案放在編譯器可以找到的地方，例如放在 TcnsorFlow 原始檔案夾的根目錄下。

（3）使用 SELECTIVE_REGISTRATION 建置 TensorFlow，例如將 --copts="-DSELECTIVE_REGISTRATION" 傳遞給 Bazel 建置指令。

此過程會重新編譯函數庫，以便只包含所需操作和類型，這樣可以顯著減少可執行檔的大小，我們在開發中一定要有這個步驟。舉例來說，在 Inception v3 中，新的模型檔案大小只有 1.5MB。

8.2.6 重新訓練行動資料

在行動應用上執行模型時，產生精度問題的最大原因是，使用了沒有代表性的訓練資料。舉例來說，大多數的 Imagenet 照片都是精心設計的，所以物體位於照片的中心，光線充足，鏡頭正常。

行動裝置拍攝的照片尤其自拍照，通常照相環境很差，照明不足，可能會造成魚眼失真。解決方案是，擴充開發者應用程式中實際收集的資料。這一步可能需要額外的工作，因為開發者必須要自己進行標記。

雖然我們只是擴充了原始的訓練資料，但是相當大地加強了準確性。這比改變模型架構或使用不同的技術要有效得多。

8.2.7 最佳化模型載入

大多數作業系統允許開發者使用記憶體對映載入檔案，而非透過通常的輸入輸出 API。我們不希望在記憶體堆積上分配一個記憶體區域，將模型位元組從磁碟複製到記憶體中，我們希望僅告訴作業系統檔案的內容，讓作業系統把模型對映到記憶體中。這種方法的優點是，作業系統知道整個檔案將立即被讀取，並且可以有效地規劃載入過程，因此可以非常快。實際的載入也可以暫停，直到記憶體被第一次存取，所以它與程式的初始化是非同步的。

開發者也可以告訴作業系統，你只能從記憶體區域讀取資料，而非寫入資料。這樣做的好處是，當 RAM 受到儲存壓力時，系統不會把資料從虛擬記憶體裡儲存到磁碟，而是可以放棄，因為磁碟上已經有了模型，我們可以再次載入，節省了大量的磁碟寫入時間。

由於 TensorFlow 模型通常可能有幾百萬位元組或更大，因此加快載入過程對行動和嵌入式應用程式來説是一個很大的幫助，減少寫入負載也可以對系統回應性有很大的幫助。

減少記憶體使用量也是非常有用的。舉例來説，在 iOS 上，系統可以終止使用超過 100MB RAM 的應用程式，特別是在較舊的裝置上。記憶體對映檔案使用的記憶體不會計入這個限制，所以它通常是這些裝置上模型的最佳選擇。

組成 TensorFlow 模型的大部分是模型的加權。由於 Protobuf 序列化格式的限制，我們必須對模型載入和處理程式進行一些更改。記憶體對映的工作方式是，假如我們有一個單獨的檔案，其中第一部分是一個正常的 GraphDef，它序列化成協定緩衝區格式，加權可以透過直接對映的形式載入。要建立此檔案，讀者需要執行 tensorflow/contrib/ 目錄下的 util:convert_graphdef_memmapped_format 工具。該工具接受一個已經透過 freeze_graph 執行的 GraphDef 檔案，並將其轉為最後附加加權的格式。由於該檔案不再是標準的 GraphDef Protobuf，因此需要對載入程式進行一些更改。讀者可以在 iOS 相機示範應用程式 LoadMemoryMappedModel() 函數中看到這個範例。

8.2.8 保護模型檔案

在預設情況下，開發者的模型將以磁碟上的標準序列化 Protobuf 格式儲存。理論上這表示任何人都可以複製你的模型，所以我經常被問及如何防止這種情況。在實作中，大多數模型都是特定於應用程式的，並且透過最佳化來混淆和隱藏模型，可以避免競爭對手採用類似於拆解和反向工程的風險。如果你想讓臨時使用者更難存取你的檔案，可以使用下面的方法進行加密和解密。

大多數範例使用 ReadBinaryProto 簡單地從磁碟載入 GraphDef。這一步需要讀磁碟上的未加密的 Protobuf。幸運的是，呼叫的實現過程非常簡單，撰寫一個可以在記憶體中解密的相等函數是很容易的。以下是一些程式，展示了如何使用自己的解密過程來讀取和解密 Protobuf：

```
Status ReadEncryptedProto(Env* env, const string& fname,
    ::tensorflow::protobuf::MessageLite* proto) {
  string data;
  TF_RETURN_IF_ERROR(ReadFileToString(env, fname, &data));
  DecryptData(&data);

  if (!proto->ParseFromString(&data)) {
    TF_RETURN_IF_ERROR(stream->status());
    return errors::DataLoss("Can't parse ", fname,
    " as binary proto"); } return Status::OK();
}
```

要使用這段程式，開發者需要自己定義 DecryptData() 函數。它可以像下面的程式一樣簡單：

```
void DecryptData(string* data) {
    for (int i = 0; i < data.size(); ++i) {
    data[i] = data[i] ^ 0x23;
    }
}
```

上面的程式是一個概念性的展示，實際的程式會更複雜一些。

8.2.9 量化計算

神經網路中最有趣的研究領域之一就是如何降低模型精度。在預設情況下，用於計算的最方便的格式是 32 位元浮點數，但是由於大多數網路在訓練後對雜訊具有容錯性，事實證明很多推理可以在 8 位元或更少的情況下執行，準確性沒有太大的損失。我們將在 11.2.1 節中談到這個問題，介紹如何透過 quantize_weights 轉換來縮小模型的檔案大小。8 位緩衝區在用於計算之前會擴充到 32 位元浮點數，因此對網路的變化是相當小的。所有其他的操作只是看到正常的浮點輸入。

更激進的方法是，嘗試使用 8 位標記法進行盡可能多的計算。許多 CPU 都有 SIMD 指令（如 NEON 或 AVX2），它們可以在每個週期執行更多的 8 位計算。這也表示像 Qualcom 的 HVX DSP 或 Google 的張量處理單元這樣的專門硬體可能無法極佳地支援浮點運算，不能加速神經網路的計算。

從理論上講，我們沒有理由用少於 8 位的資料進行運算，實際上在很多實驗中，我們已經看到 7 位甚至 5 位都是可用的，且沒有太多的損失。然而，我們沒有足夠的硬體可以支援這些少於 8 位元的運算單元。

1 量化挑戰

量化方法所面臨的最大挑戰是，神經網路中相當隨機的數字範圍，這些數字事先是不知道的，所以將它們擬合成 8 位非常困難。建立算數運算來使用這些表示也是非常棘手的，因為我們必須重新實現在使用浮點時免費獲得的許多應用程式，例如範圍檢查。因此，最佳化後的程式與相等的浮動版本會有很大的不同。還有一個問題是，由於我們不知道什麼時候會輸入什麼，所以中間計算的範圍很難估計。

2 量化表示

因為我們會處理數值較大的陣列，這些數值通常分佈在一個共同的範圍內，所以可以使用浮點值的最小值和最大值將這些值線性地編碼為 8 位

元。實際上,這看起來很像一個區塊浮點表示,儘管實際上它更靈活一些。下面是使用原始值對浮點陣列進行編碼的範例:

```
[-10.0, 20.0, 0.0]
```

讀者看一下這個陣列,可以看到最小值和最大值分別是 –10.0 和 20.0。取陣列中的每個值分別減去最小值 –10,除以最大值和最小值之間的差值(這裡是 20-(-10)=30),獲得 0.0~1.0 之間的歸一化值,然後分別乘以 255,轉換成 8 位數值。算式如下:

```
[((-10.0 - -10.0) / 30.0) * 255, ((20.0 - -10.0) / 30.0) * 255,  ((0.0 -
-10.0) / 30.0) * 255]
```

結果如下:

```
[0, 255, 85]
```

在處理這種量化表示時,最關鍵的一點是記住如果我們不知道它所基於的最小值和最大值,這種表示就沒有意義。最好把這種方法想像成一個實數的壓縮方案,只有在最大和最小值都定義的情況下,數值才有意義。

在 TensorFlow 中,這表示每次量化張量透過圖形時,都需要確保有兩個輔助張量來儲存最小和最大浮點數。所以所有接受張量作為輸入的的緩衝器都要支援輔助張量。同時輸出的定點值也有對應的輔助變數可以獲得最大和最小值。

❸ 數值代表性的不足

這種表示的好處是非常明顯的。如果將最小值和最大值設定為 2 的冪,則可以使用它來儲存傳統的定點值,該範圍不必像典型的有號表示那樣對稱,並且可以適應幾乎任何比例值。

當我們開始使用量化的時候，這些屬性是非常重要的，因為我們沒有清楚地了解我們可以把什麼約束放在這些值上，而不會失去整體的精確性。隨著我們獲得更多的經驗，我們意識到我們可以在不明顯影響準確性的情況下進行更嚴格的限制，而且這種靈活的格式也有缺點。

在神經網路中，0 是一個特殊的數字，因為它被用於填充超出卷積矩陣的影像數值，並且是來自 Relu 啟動函數的任何負數的輸出。這就產生了一個微妙的問題，即如果 0 沒有一個確切的表示，舉例來説，如果最接近的編碼值實際解碼為 0.1 而非 0.0，那麼這個錯誤會損害網路整體的準確性。

量化在神經網路上運算時，由四捨五入引起的誤差類似他們在訓練時產生的雜訊。這表示量化誤差必須大致均勻，或在足夠大的執行中至少平均為零。如果數值編碼範圍內的每個數字出現的頻率相同，並且當 0 比任何其他數字都多得多時，量化誤差就可能被放大。這種結果會導致系統偏差，最後導致產生錯誤的結果。

另一個缺點是，可能會造成無意義的或無效的數值表示。舉例來説，在最小值和最大值相等或最小值大於最大值的情況下，這種數值表示可能顯示這種表示本身就是一種缺陷。

此外，當我們試圖用相同的數值表示來表現 32 位數字時，數值表示的範圍可能會變得非常大且很難實現。

為了解決這些問題，我們可以限制最小值和最大值可以達到的值。舉例來説，當我們定義包含 0 的範圍的數值範圍時，如果最小值和最大值太接近，那麼可以將它們分開一點。將來，我們也可以強制數值對稱分佈。幸運的是，我們能夠在不改變數值表示的情況下解決這個問題。在程式進行量化計算時，要盡可能強制這些規則，以獲得更好的計算結果。

8.2.10 使用量化計算

產生定點模型的方法通常是，採用一個經過浮點數訓練的模型，使用轉換工具進行浮點到定點的轉換。這個過程要盡可能用量化運算取代浮點數運算。由於 8 位元演算法的實現與浮點數完全不同，所以我們把轉換的重點放在流行模型中常用的操作上，以下是最新的 8 位操作的列表：

- BiasAdd
- Concat
- Conv2D
- MatMul
- Relu
- Relu6
- AvgPool
- MaxPool
- Mul

這些操作足以實現 Inception 模型和許多其他卷積模型。在這種網路結構中，相鄰 Ops 之間，我們使用 8 位量化的緩衝區來傳遞數值。如果遇到不支援的操作，任何量化的張量都將被轉為浮點數，並以正常的方式執行，在下一個 8 位操作之前再轉換回量化計算。

我們來看一個 Relu 操作的實例。所有的 Relu 所做的是取一個張量陣列並輸出其輸入的拷貝，但任何負數都被 0 代替。量化這個操作所做的第一件事是，用相等的 8 位版本（稱為量化的 Relu）取代 Relu。此時，我們暫不關心和它相關的操作，它的輸入和輸出仍是浮點數。

我們使用量化操作來處理輸入，這個操作輸出浮點最小值和最大值，以及 8 位的編碼值。然後由 QuantizedRelu 進行計算操作，它將 8 位值與範圍一起輸出。實際上，對於這個實現，輸出範圍與輸入範圍相同，所以我們可以直接將來自量化的最小值和最大值作為輸入來量化，但是為了保持執行一致，對於每個量化輸出總有一個約定的數值範圍。

QuantizedRelu 編碼的 8 位值與數值範圍一起被送入 Dequantize 操作，產生最後的浮點輸出，這種處理方式非常複雜。重要的是，這是一種通用的方法，我們先把浮點數量化並轉換成定點數，進行定點數計算，然後把定點數結果轉換成浮點數。我們可以只關注其中的 8 位的計算操作，而不必了解相關節點的運算。

一般來說我們會檢查一遍 TensorFlow 的計算圖，把運算做一次取代，然後透過結果圖去除效率不佳的操作。透過發現這種模式，轉換工具可以刪除那些不必要的操作，產生簡化的計算圖。

1 去量化的重要性

消除量化計算非常重要，因為這表示 8 位元運算圖的效能取決於模型中有多少操作具有量化的相等性，如果未轉換的運算元在浮點和 8 位之間引起大量的轉換，那麼在運算圖的開頭或結尾處進行一些轉換不會太重要。舉例來說，SoftMax 通常是工作量很小的操作，並作為運算圖的最後一步，所以它通常不是一個瓶頸。但是，如果再在運算圖的核心部分混合使用浮點和 8 位進行計算，會對數值低位上的精確度造成影響，就失去了使用 8 位元定點數的意義。

需要注意的問題是，要有效地執行 8 位元演算法需要系統特定的 SIMD 程式的支援。我們使用 gemmlowp 函數庫來實現組成大量神經網路計算的矩陣乘法，但是目前只針對 ARM NEON 和行動 Intel 晶片進行了最佳化。實作中，我們使用 Eigen 針對這些晶片進行了高度最佳化的浮點函數庫來加強計算品質。

2 啟動範圍

上面還沒有提到的挑戰是,一些採用 8 位輸入的操作實際上產生了 32 位元輸出。舉例來說,如果開發者正在進行矩陣乘法,則每個輸出值將是一系列相互相乘的 8 位輸入數字的和。乘以 2 個 8 位輸入的結果是一個 16 位元的值,為了準確地累加它們中的一些需要大於 16 位元的資料,這在大多數晶片上表示要使用一個 32 位元數值的計算單元。然而,使用這個結果作為輸入的後續量化操作並不需要 32 位元資料,因為對後續操作強制進行浮點數運算並不一定更有效,而且處理起來也相當困難。相反,我們通常將這 32 位元轉為 8 位數進行運算。

可以想像計算一個特定矩陣乘法產生的最小和最大的可能值,並把這些值看成較大範圍數值中的可以用 8 位數表示的數值。然而,這種編碼是比較低效的,因為大多數神經網路操作的實際輸入不具有極端的分佈,所以我們實作中遇到的最小和最大的值將比他們的理論極限小得多。使用極端最大和最小值的範圍表示實際計算中的大部分數值將被浪費。

為了解決這個問題,我們需要知道通常遇到的資料極端情況。不幸的是,這些資訊已經被證明是非常難以計算和分析的,所以我們最後不得不在實際的測試中檢驗和觀察統計資料,我們只能透過執行整個神經網路來解決這個問題。常用的處理方法有以下三種。

(1)動態範圍
確定動態值範圍最簡單的方法是,在產生一個 32 位元操作之後,插入一個運算操作來計算出這些值的實際範圍。然後可以將這個範圍輸入一個重新量化的操作中,該操作將 32 位元張量轉為 8 位元,並指定一個目標輸出範圍。

這種方法的一大優點是，不需要額外的資料或使用者干預，所以這是 quantize_nodes 轉換使用的預設方式。這種方法使得浮動網路變得簡單，並將其轉為 8 位，然後可以檢查精度和效能。

這種方法的缺點是，每次執行推斷都需要執行範圍計算，即檢視每個輸出值並計算每個緩衝區的最小值和最大值。這是額外的工作，在 CPU 上會引起效能降低，雖然這不是嚴重的效能損失，但是對某些專用的機器學習的硬體平台來說，由於這些硬體可能無法處理這種動態的數值轉換，所以會消耗額外的運算資源而引起嚴重的效能損失。。

（2）觀察值範圍

確定觀察值範圍最簡單的方法是，透過網路執行一組具有代表性的資料，追蹤每個操作的範圍隨著時間的演進，使用統計方法來估計合理的值來覆蓋所有這些範圍，而不會浪費太多的精度。不幸的是，這種方法很難自動完成，因為組成代表性資料的想法依賴於應用程式。

舉例來說，對於影像識別網路，以亂數或極限值作為輸入的情況下，只會使用少數的模式識別路徑，所以獲得的範圍將不是很有用。相反，需要使用具有代表性的資料，例如訓練資料。

為了做到這一點，TensorFlow 提供了一個多階段的方法，即使用 insert_logging 規則增加每次執行模型時都會輸出範圍值的偵錯操作。接下來我們示範一個完整的在預訓練的 InceptionV3 上實現的實例。

首先，下載並解壓縮模型檔案：

```
mkdir /tmp/model/
curl "https://storage.googleapis.com/download.tensorflow.org/ \
      models/inception_dec_2015.zip" \
      -o /tmp/model/inception_dec_2015.zip
      unzip /tmp/model/inception_dec_2015.zip -d /tmp/model/
```

然後，使用 8 位計算進行量化：

```
bazel build tensorflow/tools/graph_transforms:transform_graph
bazel-bin/tensorflow/tools/graph_transforms/transform_graph \
    --in_graph="/tmp/model/tensorflow_inception_graph.pb" \
    --out_graph="/tmp/model/quantized_graph.pb" --inputs='Mul:0' \
    --outputs='softmax:0' --transforms='add_default_attributes
strip_unused_nodes(type=float, shape="1,299,299,3")
remove_nodes(op=Identity, op=CheckNumerics)
fold_old_batch_norms
quantize_weights
quantize_nodes
Strip_unused_nodes'
```

一旦完成，執行 label_image 範例以確保模型仍然提供預期的結果。它預設執行在 Grace Hopper 的圖片上，所以 Uniform 是最好的輸出結果：

```
bazel build tensorflow/examples/label_image:label_image
bazel-bin/tensorflow/examples/label_image/label_image \
    --input_mean=128 --input_std=128 --input_layer=Mul \
    --output_layer=softmax --graph=/tmp/model/quantized_graph.pb \
    --labels=/tmp/model/imagenet_comp_graph_label_strings.txt
```

接下來，我們把記錄檔操作附加到所有 RequantizationRange 節點的輸出上：

```
bazel-bin/tensorflow/tools/graph_transforms/transform_graph \
    --in_graph=/tmp/model/quantized_graph.pb \
    --out_graph=/tmp/model/logged_quantized_graph.pb \
    --inputs=Mul \
    --outputs=softmax \
    --transforms='insert_logging(op=RequantizationRange, \
    show_name=true, message="__requant_min_max:")'
```

現在可以執行該圖了，**stderr** 應該包含記錄檔記錄，顯示該範圍在該執行時期的值。

```
bazel-bin/tensorflow/examples/label_image/label_image \
    --input_mean=128 --input_std=128 \
    --input_layer=Mul --output_layer=softmax \
    --graph=/tmp/model/logged_quantized_graph.pb \
    --labels=/tmp/model/imagenet_comp_graph_label_strings.txt 2> \
    /tmp/model/logged_ranges.txt
cat /tmp/model/logged_ranges.txt
```

我們會看到一系列這樣的行輸出：

```
;conv/Conv2D/eightbit/requant_range__print*; *requant_min_max:
[-20.887871] [22.274715]
```

輸出中，我們可以獲得這些運算元值的範圍值。這裡只是在單一影像上執行一次計算圖形。在實際應用中，我們希望執行數百個有代表性的影像，以獲得所有常用數值的範圍。

最後，使用 freeze_requantization_ranges 轉換來取得我們收集的資訊，並用簡單的常數代替動態範圍計算，程式如下：

```
bazel-bin/tensorflow/tools/graph_transforms/transform_graph \
    --in_graph=/tmp/model/quantized_graph.pb \
    --out_graph=/tmp/model/ranged_quantized_graph.pb \
    --inputs=Mul \
    --outputs=softmax \
    --transforms='freeze_requantization_ranges \
    (min_max_log_file=/tmp/model/logged_ranges.txt)'
```

如果在這個新的網路上執行 label_image 實例，應該能夠看到一致的結果，儘管確切的計算數字可能與以前有些不同：

```
bazel-bin/tensorflow/examples/label_image/label_image \
    --input_mean=128 --input_std=128 --input_layer=Mul \
    --output_layer=softmax \
    --graph=/tmp/model/ranged_quantized_graph.pb \
    --labels=/tmp/model/imagenet_comp_graph_label_strings.txt
```

（3）訓練值範圍
關於訓練值範圍，我們以計算啟動層數值的方法來舉例。

計算啟動層數值範圍的方法是，在訓練過程中發現數值範圍。開發者可以使用 FakeQuantWithMinMaxVars 的 Ops 來執行此操作。這個 Ops 操作可以放在圖中的不同點上，在由兩個代表最小值和最大值的變數輸入設定的範圍內，透過將浮點數的輸入值轉換成一個定點數的方法來模擬量化不準確性。這些輸入值範圍會在更新的過程中根據設定的最大值、最小值而變化。

不幸的是，在 Python 中沒有任何方便的函數可以將這些操作增加到你的模型中，所以如果讀者採用這種方法，就要在特定的範圍內手動插入操作和變數。

這種方法有兩個好處，可以彌補它在使用中的不便。當我們採用預訓練的浮點數模型並簡單地將其轉為 8 位時，精度通常會有小幅度的下降，舉例來説，InceptionV3 上的 top-1 精度從 78％下降到 77％。但如果把量化整合進訓練通常會大幅縮小損失，或事先知道數值範圍也有助減少推斷過程中的延遲。

8.3 設計機器學習應用程式要點

在設計機器學習的應用時，我們要注意以下兩點：

（1）如果必須使用人工智慧，我們的目標是什麼？我們必須要問自己這個問題。請注意以上我們討論過的這些問題，並且開發者可能還要處理許多行動裝置的特定問題。因此請確定機器學習是正確的且是唯一的解決方案。

以規則為基礎的解決方案可以解決一些問題。如果開發者一定要在問題上應用人工智慧，可以先嘗試一個簡單的解決方案或進行更多離線測試，再開始在行動裝置上工作。

開發者要選擇正確的機器學習模型和方式。舉例來說，開發者只需要進行推理，或要同時進行推理和訓練。選擇正確且恰當的機器學習方式，會相當大減輕開發的難度。

（2）嘗試盡可能減少應用的大小。大型應用通常表示更大的記憶體需求，回應慢、低效能和糟糕的使用者體驗。我們需要確定應用增加的部分是由應用程式或機器學習引起的，通常機器學習可以因為引用大模型的檔案而增大。我們可以透過設定 "SELECTIVE_REGISTRATION" 刪除模型檔案中的不必要的操作。

如果開發者在應用中嵌入機器學習模型，使用者將首先體驗到慢速下載。解決方法是，使用者在第一次執行應用程式時下載模型。在這種情況下，如果網路連接不穩定或足夠快，開發者還需要準備一個後備計畫。後備計畫可以是以規則為基礎的解決方案，也可以是非常小的機器學習解決方案。

在 Android 上一般使用 Java API。透過使用原生 C/C++ API，開發者可能會增加應用大小，並且開發者必須為不同的平台 / 硬體建置原生程式。在測試中，我們發現呼叫 Java API 不一定會使應用程式變慢，效能的瓶頸通常是執行模型。

同時，我們還要考慮一些其他因素：

■ 重複使用記憶體碎片。

■ 模型可以儲存在資源檔案夾或暫存檔案夾中，Java 有 API 支援載入模型檔案，使用者不必使用原始 API 載入模型。

■ 必要時使用原生 API OpenGL。

■ 透過硬體加速或運算量化來提高性能。

■ 尊重 Android 執行規則，儘量節省電池。

■ 保護模型。一般來說機器學習執行機制不知道執行圖中的上下文，只需遵循應用程式列出的任何指令。如果模型被修改或遭駭客攻擊，則輸出結果一定不是我們所期望的。如果使用下載模型，則可能需要在執行模型之前進行健全性檢查。

■ 防止模型濫用。如果模型和應用是分開的，開發者可能希望保護它免受濫用。而當模型在相對安全的資料中心執行時期，開發者可能不會遇到這種挑戰。

TensorFlow 的硬體加速

本章介紹如何利用現有的行動裝置建置和執行一個機器學習模型，以及如何使用硬體加速。現在主流的行動裝置中，有兩種裝置支援硬體加速，華為和高通 Qualcomm。下面介紹如何在這兩種裝置上開發和執行機器學習的應用。

9.1 神經網路介面

Android 神經網路介面（Android Neural Networks API，簡稱 NNAPI）是一個 Android C API，專門為在行動裝置上執行機器學習的密集型運算而設計的 API。NNAPI 主要在為建置和訓練神經網路的更進階機器學習架構（例如 TensorFlow Lite、CAFFE2 或其他）提供一個基礎的功能層。API 適用於執行 Android 8.1（API 等級為 27）或更新版本的所有裝置。

NNAPI 推理（Inference）的過程是，使用開發者已訓練的自訂模型，從 Android 裝置中讀取資料、執行模型並獲得結果。一些典型的推理的應用實例包含影像分類、預測使用者行為，以及選擇對搜索查詢的適當回應等。

在行動裝置上推理具有以下優勢：

- 延遲時間：不需要透過網路連接發送請求並等待回應。這對處理從攝影機傳入的連續畫面的視訊應用非常重要。
- 可用性：應用甚至可以在沒有網路覆蓋的條件下執行。
- 速度：與單純的通用 CPU 相比，特定於神經網路處理的新硬體可以提供顯著加快的計算速度。
- 隱私：資料不會離開裝置。
- 費用：所有計算都在裝置上執行，不需要額外伺服器。

但是，開發者也應考慮它一些的利弊：

- 系統使用率：神經網路封包含很多計算，這會增加電池消耗。開發者的應用需要注意耗電量，應當考慮監視電池的執行狀況，尤其要針對長時間執行的計算進行監視。
- 應用大小：注意模型的大小。模型可能會佔用很多空間。由於在開發者的 APK 中綁定較大的模型會過度地影響使用者，所以開發者需要考慮在應用安裝後下載模型、使用較小的模型，或在雲中執行計算。但 NNAPI 未提供在雲中執行模型的功能。

9.1.1 了解 Neural Networks API 執行時期

NNAPI 的目的是被機器學習程式、架構和工具呼叫，這些可以讓開發者脫離物理裝置，而訓練他們的模型並將模型部署在 Android 裝置上。

Android 的應用一般不會直接使用 NNAPI，但會使用更進階的機器學習架構。這些架構反過來可以使用 NNAPI 在被支援的裝置上執行硬體加速的推理運算。

根據應用的要求和裝置上的硬體能力，Android 的神經網路執行時期（Runtime）可以在可用的處理器（包含專用的神經網路硬體、圖形處理單元（GPU）和數位訊號處理器（DSP）之間有效地分配計算工作負載。對於缺少專用的供應商驅動程式的裝置，NNAPI 執行時期（Runtime）將在 CPU 上執行最佳化後的程式。NNAPI 的系統架構如圖 9-1 所示。

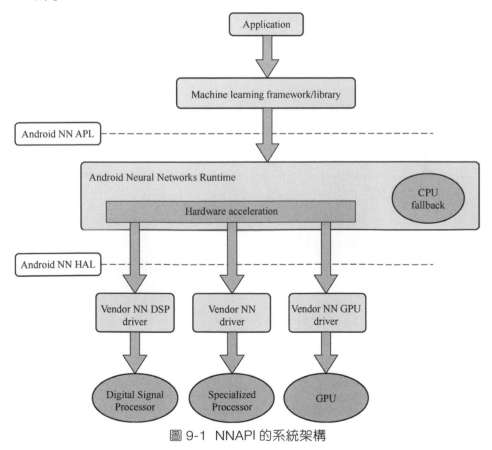

圖 9-1　NNAPI 的系統架構

9.1.2 Neural Networks API 程式設計模型

要使用 NNAPI 執行計算，開發者首先需要建置一個可以定義要執行的計算的有向圖（Directed Graph）。此計算圖與輸入資料（例如從機器學習架構傳遞過來的加權和偏差）相結合，組成 NNAPI 執行時期評估的模型。

NNAPI 定義了以下四個概念：

模型：是數學運算和透過訓練過程學習的常數值的計算圖。這些運算特定於神經網路。它們包含二維（2D）卷積、邏輯（Sigmoid）啟動和全連接（ReLU）啟動等。建立模型是一個同步操作，一旦建立成功就可以在執行緒和編譯之間重用模型。在 NNAPI 中，一個模型表示為一個 ANeuralNetworksModel 實例。

編譯：表示用於將 NNAPI 模型編譯到更低階的程式中。建立編譯是一個同步操作，一旦成功建立就可以在執行緒和執行之間重用編譯。在 NNAPI 中，每個編譯表示為一個 ANeuralNetworksCompilation 實例。

記憶體：表示共用記憶體、記憶體對映檔案和類似的記憶體緩衝區。使用記憶體緩衝區可以讓 NNAPI 在執行時期將資料更高效率地傳輸到驅動程式中。一個應用通常會建立一個共用記憶體緩衝區，其中包含定義模型所需的每一個張量。也可以使用記憶體緩衝區來儲存執行實例的輸入和輸出。在 NNAPI 中，每個記憶體緩衝區表示為一個 ANeuralNetworksMemory 實例。

執行：用於將 NNAPI 模型應用到一組輸入並擷取結果的介面。執行是一種非同步作業。多個執行緒可以在相同的執行上等待。當執行完成時，所有的執行緒都將釋放。在 NNAPI 中，每一個執行表示一個 ANeuralNetworksExecution 實例。

基本的程式設計流程如圖 9-2 所示。

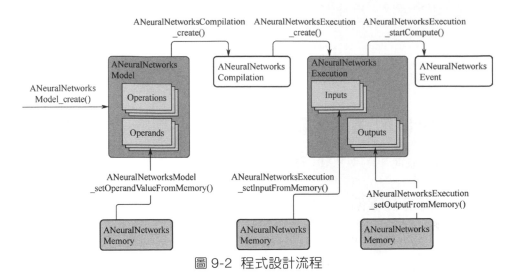

圖 9-2 程式設計流程

下面的程式是 **NNAPI** 介面的定義。請注意，現在的版本是 1.2 版本。其中的程式註釋被刪除了，只保留程式的核心部分。

```
package android.hardware.neuralnetworks@1.2;

import @1.0::ErrorStatus;
import @1.0::IPreparedModelCallback;
import @1.1::ExecutionPreference;
import @1.1::IDevice;

/**
 * This interface represents a device driver.
 */
interface IDevice extends @1.1::IDevice {
    getVersionString() generates (ErrorStatus status, string version);

    getSupportedOperations_1_2(Model model)
            generates (ErrorStatus status, vec<bool> supportedOperations);
```

```
    prepareModel_1_2(Model model, ExecutionPreference preference,
                     IPreparedModelCallback callback)
         generates (ErrorStatus status);
};
```

9.1.3 NNAPI 實現的實例

本節介紹一個簡單的 NNAPI 實現的實例。

1 下載和安裝實例

首先，下載 android-ndk.git，nn_sample 資料夾裡有簡單的 NNAPI 的實現；然後，使用 Gradle（指令為 gradlew）建置並產生 app/build/outputs/apk/debug/app-debug.apk。實際程式如下：

```
$ git clone https://github.com/googlesamples/android-ndk.git
$ cd android-ndk/nn_sample
$ ./gradlew build
```

由於這個實例需要編譯原生 C 程式，實例裡使用了 cmake，所以要安裝 cmake 和 ninja。在 Ubuntu 上可以透過執行指令 "sudo apt-get cmake" 來安裝 cmake，安裝 ninja 可到 https://github.com/ninja-build/ninja/releases 下載。

除了使用 Gradle，我們還嘗試使用 Bazel 去編譯這個應用，看看有什麼不同。使用 Bazel 之前，要新增 workspace 和 build 檔案，TensorFlow Lite 已經把這些都做好了，我們可以借用一下。把 nn_sample 資料夾拷貝到 tensorflow/examples 下，在 tensorflow/examples/ nn_sample/app/src/main 下新增一個 build 檔案，把下面的內容複製到 build 檔案中：

```
package(default_visibility = ["//visibility:public"])

licenses(["notice"])  # Apache 2.0

load(
    "//tensorflow:tensorflow.bzl",
    "tf_copts",
    "tf_opts_nortti_if_android",
)

exports_files(["LICENSE"])

LINKER_SCRIPT = "//tensorflow/contrib/android:jni/version_script.lds"

cc_binary(
    name = "libnn_sample.so",
    srcs = glob([
        "cpp/**/*.cpp",
        "cpp/**/*.h",
    ]),
    copts = tf_copts(),
    defines = ["STANDALONE_DEMO_LIB"],
    linkopts = [
        "-landroid",
        "-ldl",
        "-ljnigraphics",
        "-llog",
        "-lm",
        "-lneuralnetworks",
        "-z defs",
        "-s",
        "-Wl,--version-script",  # This line must be directly followed by
LINKER_SCRIPT.
```

```
        "$(location {})".format(LINKER_SCRIPT),
    ],
    linkshared = 1,
    linkstatic = 1,
    tags = [
        "manual",
        "notap",
    ],
    deps = [
        LINKER_SCRIPT,
    ],
)

cc_library(
    name = "libnn_sample",
    srcs = [
        "libnn_sample.so",
    ],
    tags = [
        "manual",
        "notap",
    ],
)

android_binary(
    name = "nnapi_demo",
    srcs = glob([
        "java/**/*.java",
    ]),
    aapt_version = "aapt",
    assets = [
        ":model_assets",
    ],
    assets_dir = "assets",
```

```
    custom_package = "com.example.android.nnapidemo",
    inline_constants = 1,
    manifest = "AndroidManifest.xml",
    nocompress_extensions = [
        ".bin",
    ],
    resource_files = glob(["res/**"]),
    tags = [
        "manual",
        "notap",
    ],
    deps = [
        ":libnn_sample",
        "@androidsdk//com.android.support.constraint:constraint-layout-1.0.2",
    ],
)


filegroup(
    name = "model_assets",
    srcs = [
        "assets/model_data.bin",
    ],
)
```

Gradle 工具的使用依賴於 'com.android.support.constraint:constraint-layout:1.1.3'，原生的 Android NDK 不附帶這個版本，但是由於不影響計算結果，所以這裡直接使用 "@androidsdk//com.android.support.constraint:constraint-layout-1.0.2" 來替代 1.1.3 版本，讀者也可以從 maven 下載使用最新版本。

這裡還要指定 nocompress_extensions 參數禁止壓縮帶有 ".bin" 的檔案。使用 cc_binary 建置二進位檔案 libnn_sample.so，使用 cc_library 建置動

態連結程式庫。現在，執行下面的指令：

```
$ bazel build --fat_apk_cpu=arm64-v8a --android_cpu=arm64-v8a
//tensorflow/examples/nn_sample/app/src/main:nnapi_demo
```

產生檔案 nnapi_demo.apk，使用指令 "unzip –l" 去檢視檔案內容，APK 的 assets 裡應該包含 model_data.bin。讀者可以將 nnapi_demo.apk 檔案和 Gradle 建置的檔案進行比較：

```
$ unzip -l bazel-bin/tensorflow/examples/nn_sample/app/src/main/nnapi_demo.apk
  Length      Date     Time    Name
---------  ----------  -----    ----
   194720  2010-01-01 00:00    lib/arm64-v8a/libnn_sample.so

$ unzip -l ./app/build/outputs/apk/release/app-release-unsigned.apk
  Length      Date     Time    Name
---------  ----------  -----    ----
   223312  1980-00-00 00:00    lib/arm64-v8a/libnn_sample.so
```

Bazel 建置的檔案大小少於 200KB，Gradle 建置的大於 200KB。同樣的技術也可以用於 Gradle 去減少 APK 的檔案大小，只是需要改進 Gradle 的建置過程，而這些 TensorFlow 裡的 build 已經都做好了。

2 模型準備

這個實例示範了一個簡單的 NNAPI 的實現。程式入口是 nn_sample/app/src/main/ java/com/example/android/nnapidemo/MainActivity.java。 在 onCreate 裡載入模型檔案 model_data.bin，程式如下：

```
extern "C"
JNIEXPORT jlong
JNICALL
Java_com_example_android_nnapidemo_MainActivity_initModel(
```

```
        JNIEnv *env,
        jobject /* this */,
        jobject _assetManager,
        jstring _assetName) {

// 取得模類型資料檔案的檔案描述符號
    AAssetManager *assetManager = AAssetManager_fromJava(env, _assetManager);
    const char *assetName = env->GetStringUTFChars(_assetName, NULL);
    AAsset *asset = AAssetManager_open(assetManager, assetName,
AASSET_MODE_BUFFER);
    if(asset == nullptr) {
        __android_log_print(ANDROID_LOG_ERROR, LOG_TAG, "Failed to open the
asset.");
        return 0;
    }
    env->ReleaseStringUTFChars(_assetName, assetName);
    off_t offset, length;
    int fd = AAsset_openFileDescriptor(asset, &offset, &length);
    AAsset_close(asset);
    if (fd < 0) {
        __android_log_print(ANDROID_LOG_ERROR, LOG_TAG,
                            "Failed to open the model_data file descriptor.");
        return 0;
    }
    SimpleModel* nn_model = new SimpleModel(length, PROT_READ, fd, offset);
    if (!nn_model->CreateCompiledModel()) {
        __android_log_print(ANDROID_LOG_ERROR, LOG_TAG,
                            "Failed to prepare the model.");
        return 0;
    }

    return (jlong)(uintptr_t)nn_model;
}
```

從上面的程式中可以看到，我們使用 Android NDK 的 API AAssetManager_
open 和 AAsset_openFileDescriptor 讀取模型檔案，建置一個 SimpleModel
類別。此應用先呼叫 Java_com_example_android_nnapidemo_MainActivity_
startCompute，再呼叫 SimpleModel 類別的 Compute 去實現 tensor 的運算。

這個應用的實現非常簡單，就是 (tensor0 + tensor1) ×(tensor2 + tensor3)，
其中 tensor0 和 tensor2 是常數，是從模型檔案 model_data.bin 中讀取的，
它們在實際的應用中被看作訓練後的模型加權（Weights）。

```
        (tensor0 + tensor1) * (tensor2 + tensor3) 的圖形表示如下：
tensor0 ---+
          +--- ADD ---> intermediateOutput0 ---+
tensor1 ---+                                    |
                                           +--- MUL---> output
tensor2 ---+                                    |
          +--- ADD ---> intermediateOutput1 ---+
tensor3 ---+
```

上面這個圖形表示可以看作一個小模型，在這個模型裡一共有 8 個運算
元，分別為：

■ 2 個張量，分別供給兩個加法。

■ 2 個常數張量，分別供給兩個加法。

■ 1 個混合啟動運算元，被用於加法和乘法。

■ 2 個中間結果張量，分別代表了兩個加法的結果。這兩個張量也供給
乘法。

■ 1 個乘法的結果。

TensorFlow 的兩個重要的概念就是模型和張量，SimpleModel 類別也是
圍繞這兩個概念進行設計的。下面是 bool SimpleModel::CreateCompiled

Model() 函數的實現程式：

```
// 建立 AneralNetworksModel 控制碼
 status = ANeuralNetworksModel_create(&model_);
 if (status != ANEURALNETWORKS_NO_ERROR) {
    __android_log_print(ANDROID_LOG_ERROR, LOG_TAG, "ANeuralNetworksModel_
create failed");
    return false;
 }
```

上面程式產生一個空的模型，然後，向這個模型裡填裝運算元：

```
// 為張量增加運算元
status = ANeuralNetworksModel_addOperand(model_, &float32TensorType);
uint32_t tensor0 = opIdx++;
if (status != ANEURALNETWORKS_NO_ERROR) {
    __android_log_print(ANDROID_LOG_ERROR, LOG_TAG,
                        "ANeuralNetworksModel_addOperand failed for
                        operand (%d)", tensor0);
    return false;
}
```

上面程式實現了在模型裡產生運算數。在產生了所有必要的運算數之後，接下來產生運算子：

```
// 增加第一個 ADD operation
std::vector<uint32_t> add1InputOperands = {
     tensor0,
     tensor1,
     fusedActivationFuncNone,
};
status = ANeuralNetworksModel_addOperation(model_, ANEURALNETWORKS_ ADD,
                                    add1InputOperands.size(),
                                    add1InputOperands.data(),
                                    1, &intermediateOutput0);
```

```
    if (status != ANEURALNETWORKS_NO_ERROR) {
        __android_log_print(ANDROID_LOG_ERROR, LOG_TAG,
                        "ANeuralNetworksModel_addOperation failed for
                        ADD_1");
        return false;
    }
```

上面程式在模型裡增加了一個加法,加法的輸入運算數一個是 tensor0,
另一個是 tensor1,加法的結果是 intermediateOutput0。用這種方法,程
式把所有的運算子和連結的運算數都填充進模型,然後告訴模型輸入和
輸出的張量。實作方式程式如下:

```
// 識別模型的輸入和輸出張量
    // Inputs: {tensor1, tensor3}
    // Outputs: {multiplierOutput}
    std::vector<uint32_t> modelInputOperands = {
            tensor1, tensor3,
    };
    status = ANeuralNetworksModel_identifyInputsAndOutputs(model_,

modelInputOperands.size(),

modelInputOperands.data(),

                                            1,
                                            &multiplierOutput);
    if (status != ANEURALNETWORKS_NO_ERROR) {
        __android_log_print(ANDROID_LOG_ERROR, LOG_TAG,
                        "ANeuralNetworksModel_identifyInputsAndOutputs
failed");
        return false;
    }
```

從上面程式可以看到,使用者需輸入 tensor1 和 tensor3 的數值,tensor0

和 tensor2 是常數，multiplierOutput 是模型的輸出。然後，把已產生的 model 轉換成 ANeuralNetworksCompilation，程式如下：

```
// 建立 AneralNetworksCompetition 物件
status = ANeuralNetworksCompilation_create(model_, &compilation_);
    if (status != ANEURALNETWORKS_NO_ERROR) {
        __android_log_print(ANDROID_LOG_ERROR, LOG_TAG,
                            "ANeuralNetworksCompilation_create failed");
        return false;
    }

    // 編譯完成
    status = ANeuralNetworksCompilation_finish(compilation_);
    if (status != ANEURALNETWORKS_NO_ERROR) {
        __android_log_print(ANDROID_LOG_ERROR, LOG_TAG,
                            "ANeuralNetworksCompilation_finish failed");
        return false;
    }
```

至此，模型的準備基本完成了。

3 執行模型

首先，產生一個執行器，程式如下：

```
// 從已編譯的模型中建立 AneralNetworksExecution 物件
    // Note:
    //    1. 所有輸入和輸出資料都綁定到 AneralNetworksExecution 物件
    //    2. 可以從同一編譯模型建立多個平行處理執行實例
    // 此範例只使用已編譯模型的一次執行
    ANeuralNetworksExecution *execution;
    int32_t status = ANeuralNetworksExecution_create(compilation_, &execution);
    if (status != ANEURALNETWORKS_NO_ERROR) {
        __android_log_print(ANDROID_LOG_ERROR, LOG_TAG,
```

```
                              "ANeuralNetworksExecution_create failed");
        return false;
    }
```

然後，設定輸入和輸出，下面是範例程式。注意，這裡儘量使用共用記憶體以節省資料拷貝。

```
// AneralNetworksExecution_setinputfrommemory 將運算元與共用記憶體連結最小化原始
資料備份數的區域。注意：這裡的索引 "1" 表示 modelinput 列表的第二個運算元
    status = ANeuralNetworksExecution_setInputFromMemory(execution, 1, nullptr,
                                                         memoryInput2_, 0,
                                                         tensorSize_ *
sizeof(float));
    if (status != ANEURALNETWORKS_NO_ERROR) {
        __android_log_print(ANDROID_LOG_ERROR, LOG_TAG,
                        "ANeuralNetworksExecution_setInputFromMemory
failed for input2");
        return false;
    }
```

接下來，執行 ANeuralNetworksExecution_startCompute，程式如下：

```
// 開始執行模型
// 注意：這裡的執行是非同步的，並且將建立 AneralNetworksEvent 物件以監視執行狀態
ANeuralNetworksEvent *event = nullptr;
status = ANeuralNetworksExecution_startCompute(execution, &event);
if (status != ANEURALNETWORKS_NO_ERROR) {
    __android_log_print(ANDROID_LOG_ERROR, LOG_TAG,
                    "ANeuralNetworksExecution_startCompute failed");
    return false;
}

//等待執行完成，實現同步呼叫
status = ANeuralNetworksEvent_wait(event);
```

```
if (status != ANEURALNETWORKS_NO_ERROR) {
    __android_log_print(ANDROID_LOG_ERROR, LOG_TAG,
                        "ANeuralNetworksEvent_wait failed");
    return false;
}
```

由於我們要呼叫 ANeuralNetworksEvent_wait 去等待運算的結束，所以
運算要執行在不同的執行緒上。

最後，讀取共用記憶體的資料：

```
float *outputTensorPtr = reinterpret_cast<float *>(mmap(nullptr,
                                        tensorSize_ * sizeof(float),
                                        PROT_READ, MAP_SHARED,
                                        outputTensorFd_, 0));
```

另外，不要忘記清理使用過的記憶體：

```
ANeuralNetworksEvent_free(event);
ANeuralNetworksExecution_free(execution);

ANeuralNetworksCompilation_free(compilation_);
ANeuralNetworksModel_free(model_);
ANeuralNetworksMemory_free(memoryModel_);
ANeuralNetworksMemory_free(memoryInput2_);
ANeuralNetworksMemory_free(memoryOutput_);
```

9.2 硬體加速

本章主要討論與硬體加速有關的進展和知識，其中將簡要有關高通
（Qualcomm）、華為（Huawei NPU）及 ONNX 在硬體加速上的支援和相
關的技術。

很多公司都在積極開發機器學習的硬體支援，有兩家是公認的比較有影響力的。一個是華為 NPU，另一個是 Qualcomm DSP/GPU。

華為自主研發的人工智慧晶片應用在華為手機上，獲得了不錯的市場效果，目前在中國處於領先地位。高通在手機晶片研發上有很長的歷史，有很多技術累積。高通使用已有的 DSP 和 GPU 技術，也實現了不錯的機器學習硬體加速功能。

在筆者看來，兩家公司都獨自研發了對硬體加速的支援，試圖以自己的方式提供 "AI" 能力並嘗試建置他們的生態系統，雖然具有商業意義，但對開發人員來說卻很麻煩，這表示開發人員必須在不同的裝置上開發不同的應用程式。

筆者在華為裝置上做了一些初步的嘗試，本書記錄了一些實驗過程和結果。實驗需要一個支援 NPU 的裝置，筆者的裝置是華為 Mate 10，透過執行指令 "$ adb shell getprop | grep finger"，獲得以下傳回結果：

```
$ [ro.vendor.build.fingerprint]:[HUAWEI/ALP-AL00/HWALP:8.0.1/HUAWEIALP-
AL00/.../release-keys]
```

9.2.1 高通網路處理器

市場上有很多手機都採用高通晶片組，找到使用 Qualcomm 晶片組的手機並不難，但是還需要找到支援 GPU 或 DSP 的裝置，讀者可以參考下面的裝置列表：

Snapdragon Device	CPU	GPU	DSP
Qualcomm Snapdragon 845	Yes	Yes	Yes (CDSP)
Qualcomm Snapdragon 821	Yes	Yes	Yes (ADSP)
Qualcomm Snapdragon 820	Yes	Yes	Yes (ADSP)

```
| Qualcomm Snapdragon 710  | Yes      | Yes      | Yes (CDSP) |
| Qualcomm Snapdragon 660  | Yes      | Yes      | Yes (CDSP) |
| Qualcomm Snapdragon 652  | Yes      | Yes      | No         |
| Qualcomm Snapdragon 630  | Yes      | Yes      | No         |
| Qualcomm Snapdragon 636  | Yes      | Yes      | No         |
| Qualcomm Snapdragon 625  | Yes      | Yes      | No         |
| Qualcomm Snapdragon 450  | Yes      | Yes      | No         |
```

高通 Qualcomm 提供 SDK、範例程式等給開發人員。這裡我們會介紹高通神經網路處理器（Snapdragon Neural Processing Engine，簡稱 SNPE）。開發者需要登入高通開發者網頁（https://developer.qualcomm. com）去取得 SDK 等資源。高通登入頁面如圖 9-3 所示，登入以後就可以下載 SDK，查閱相關資源。

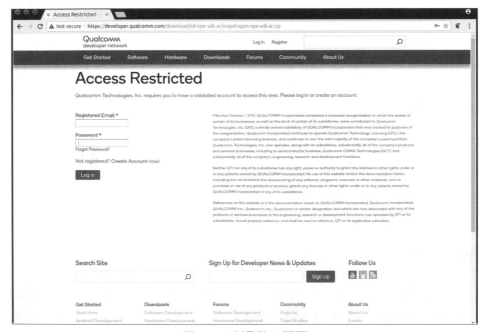

圖 9-3　高通登入頁面

Snapdragon neural Processing Engine（SNPE）的架構如圖 9-4 所示。

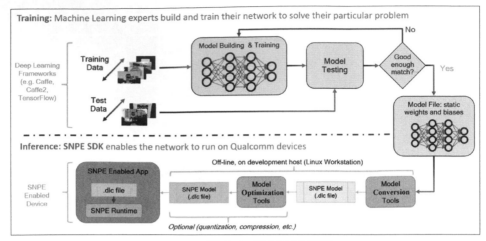

圖 9-4 SNPE 架構圖

SNPE 支 援 TensorFlow、CAFFE 和 CAFFE2，SNPE 也 支 援 ONNX。SNPE 會將 CAFFE、CAFFE2、ONNXTM 和 TensorFlow 模型轉為 SNPE 深度學習容器（DLC）檔案。讀者最好同時安裝 CAFFE 和 CAFFE2 兩個架構，因為模型是以 CAFFE2 為基礎的，並且 Android 的範例程式需要產生 alexnet dlc 檔案，該檔案將從一個指令稿產生，而這個指令稿需要 CAFFE。

筆者嘗試過市場上的幾款手機，筆者的建議是試試一款不是特別深度訂製的手機，這會減少很多開發麻煩。最後筆者選擇在小米的手機上做一些測試。測試 SNPE 之前，先執行指令 "$ snpe-net-run –h"，看到正確的執行結果說明成功，如果沒有輸出任何東西，說明失敗。如果讀者的裝置支援 DSP 或 GPU，現在就可以嘗試使用它們。

首先，從開發者網頁下載 Qualcomm Neural Processing SDK，將 snpe-1.25.1.zip 解壓後，獲得以下檔案：

```
-rw-r--r--@  1  29911  May  31  2018 LICENSE.txt
-rwxr-xr-x@  1  47090  Apr  17  03:25 NOTICE.txt
```

```
-rwxr-xr-x@  1   3869   Jun   4   10:47 REDIST.txt
-rw-r--r--@  1   7018   Jun  11   18:19 ReleaseNotes.txt
drwxr-xr-x@  4    128   Jun  30   12:11 android
drwxr-xr-x@  7    224   Jun  30   12:11 benchmarks
drwxr-xr-x@ 12    384   Jun  30   12:11 bin
drwxr-xr-x@  3     96   Jun  30   12:11 doc
drwxr-xr-x@  5    160   Jun  30   12:11 examples
drwxr-xr-x@  3     96   Jun  30   12:11 include
drwxr-xr-x@ 11    352   Jun  30   12:11 lib
drwxr-xr-x@  5    160   Jun  30   12:11 models
drwxr-xr-x@  3     96   Jun  30   12:11 share
```

現在，我們可以執行 Android 範例應用程式。這個實例程式在 examples/ android/image- classifiers 下，檔案內容如下所示：

```
drwxr-xr-x@  6    192   Jun  30   12:11 app
-rw-r--r--@  1    412   Apr  17   03:25 build.gradle
drwxr-xr-x@  3     96   Jun  30   12:11 gradle
-rw-r--r--@  1    182   Apr  17   03:25 gradle.properties
-rwxr-xr-x@  1   4971   May  31    2018 gradlew
-rw-r--r--@  1   2404   May  31    2018 gradlew.bat
-rw-r--r--@  1    158   May  31    2018 settings.gradle
-rw-r--r--@  1    735   Apr  17   03:25 setup_alexnet.sh
-rw-r--r--@  1    894   May   7   08:24 setup_inceptionv3.sh
```

讀者可以使用 Gradle 編譯執行程式。一般來説筆者會增加 Bazel 建置檔案，所以讀者也可以透過 Bazel 建置，方法如下：

首先，新增兩個 build 檔案。

一個是 app/src/main/ 下的 build 檔案，內容如下：

```
ndroid_library(
    name = "demo_lib",
    srcs = glob([
        "java/**/*.java",
```

```
    ]),
    custom_package = "com.qualcomm.qti.snpe.imageclassifiers",
    manifest = "AndroidManifest.xml",
    resource_files = glob(["res/**"]),
    deps = [
        "//app/libs:snpe_release_aar",
        "//third_party:android_arch_core_common",
        "//third_party:android_arch_lifecycle_common",
        "//third_party:com_android_support_constraint_constraint_layout_
solver",
        "//third_party:com_android_support_support_annotations",
        "@android_arch_lifecycle_runtime//aar",
        "@com_android_support_appcompat_v7//aar",
        "@com_android_support_cardview//aar",
        "@com_android_support_constraint_constraint_layout//aar",
        "@com_android_support_design//aar",
        "@com_android_support_recyclerview//aar",
        "@com_android_support_support_compat//aar",
        "@com_android_support_support_core_ui//aar",
        "@com_android_support_support_core_utils//aar",
        "@com_android_support_support_fragment//aar",
        "@com_android_support_support_v4//aar",
        "@com_android_support_support_vector_drawable//aar",
    ],
)

android_binary(
    name = "demo",
    custom_package = "com.qualcomm.qti.snpe.imageclassifiers",
    manifest = "AndroidManifest.xml",
    manifest_values = {
        "applicationId": "com.qualcomm.qti.snpe.imageclassifiers",
    },
    multidex = "native",
    resource_files = glob(["res/**"]),
```

```
    deps = [
        ":demo_lib",
    ],
)
```

另外一個是 app/libs/ 下的 build 檔案，檔案內容如下：

```
package(default_visibility = ["//visibility:public"])

aar_import(
    name = "snpe_release_aar",
    aar = "snpe-release.aar",
)
```

編譯執行檔案：

```
$ bazel build app/src/main:demo
```

執行上面的指令後獲得 Android 應用程式，安裝 Android 應用，結果如圖 9-5 所示。

圖 9-5 Android 應用圖

如果 Android 應用執行在非 Snapdragon 處理程序中，將收到以下錯誤訊息：

```
06-1208:19:54.0011681216812 D SYMPHONY: Detected Symphony running as an
Android application, using logcat for all debugging output
06-1208:19:54.0011681216812 E SYMPHONY: FATAL        0 tef4d04a4 /home/host/
build/arm-android-gcc4.9/SecondParty/symphony/src/symphony/src/lib/runtime.
cc:468 runtime_init() This version of Symphony is targeted to Snapdragon(TM)
platforms
06-1208:19:54.0011681216812 E SYMPHONY: tef4d04a4 /home/host/build/arm-
android-gcc4.9/SecondParty/symphony/src/symphony/src/lib/runtime.cc:468
**********
06-1208:19:54.0011681216812 E SYMPHONY: tef4d04a4 /home/host/build/arm-
android-gcc4.9/SecondParty/symphony/src/symphony/src/lib/runtime.cc:468 -
Terminating with exit(1)
06-1208:19:54.0011681216812 E SYMPHONY: tef4d04a4 /home/host/build/arm-
android-gcc4.9/SecondParty/symphony/src/symphony/src/lib/runtime.cc:468
**********
```

以上測試是筆者在小米手機 "Qualcomm Snapdragon 660" 上做的，下面是這款手機的型號資訊：

```
$ ro.build.fingerprint：[Xiaomi/jason/jason:7.1.1/NMF26X/V9.6.1.0.NCHCNFD:
user/release-keys]
```

實例中使用 AlexNet 的影像分類器，我們可以使用不同的硬體，來測量它們的運算速度，以下是測量的結果：

```
CPU
time: 292 ms
time: 304 ms
time: 272 ms


DSP
```

```
time: 191 ms
time: 147 ms
time: 143 ms

GPU
time: 216 ms
time: 174 ms
time: 183 ms
```

從以上測試結果可以看到，DSP 表現最佳，GPU 其次。

9.2.2 華為 HiAI Engine

華為為 AI/ML 開發提供了一套開發人員 SDK。同樣，使用者必須先到開發者網站 http://developer.huawei.com 註冊華為 ID。華為開發者首頁如圖 9-6 所示。

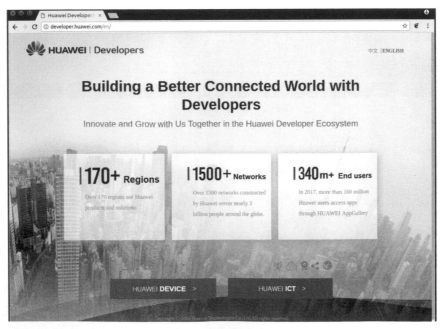

圖 9-6 華為開發者首頁

登入後，請選擇 HUAWEI DEVICE，它是行動裝置開發的入口。對個人開發人員來說，註冊非常簡單，開發者頁面如圖 9-7 所示。對於公司開發人員，註冊的流程是不一樣的，在這裡就不解釋了。

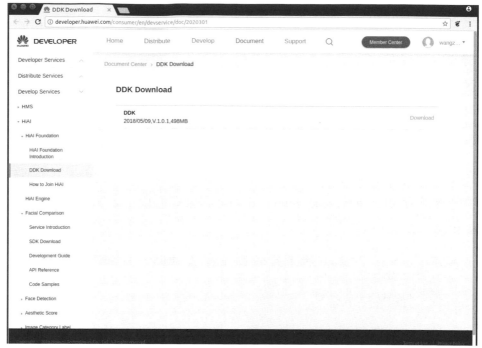

圖 9-7 華為開發者頁面

HiAI 是一個針對行動終端的人工智慧（AI）計算平台，建置了三層生態：服務能力開放、應用能力開放和晶片能力開放。整合了終端、晶片和雲的三層開放平台為使用者和開發人員帶來了更多的體驗。

華為還提供了 HiAI Engine 程式設計幫手。HiAI Engine 程式設計幫手把所有的 API 以卡片的形式呈現，卡片的詳情介紹中有針對 API 的應用場景說明、介面說明和範例程式，直接拖曳卡片到應用的專案中，就會產生 API 呼叫的 20 多行標準程式，對於更個性化的需求，開發者只需對介面中的參數稍作修改即可,比起手動輸入程式的方式,其效率要高很多,而且不容易出錯。

另外,HUAWEI DevEco IDE 還提供 7×24 小時遠端實機偵錯服務,能直接連接到華為 Openlab 實驗室的實機,實現應用功能遠端調測及應用安裝與操作。

華為提供了以下 15 個使用案例示範,這些幾乎涵蓋了目前市場中你能想到的所有使用案例。

(1) Facial Comparison

(2) Face Detection

(3) Aesthetic Score

(4) Image Category Label

(5) Image Super-Resolution

(6) Scene Detection

(7) Character Image Super-Resolution

(8) Document Correction/Detection

(9) Code Detection

(10) Face Attributes

(11) Face Orientation Recognition

(12) Face Parsing

(13) Facial Feature Detection

(14) Image Semantic Segmentation

(15) Portrait Segmentation

對於每個使用案例,開發者網站都提供了 SDK/DDK 和範例程式。SDK 一般提供進階 Java 介面。裝置開發套件(DDK)提供底層 C++ 介面。華為 HiAI 的開發流程如圖 9-8 所示。

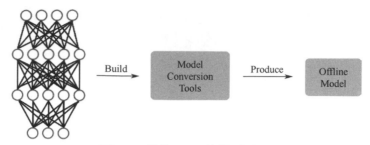

圖 9-8 華為 HiAI 的開發流程

如圖 9-9 所示說明了華為 HiAI 從 Offline 到硬體加速的開發執行流程。

圖 9-9 華為 HiAI 的開發執行流程

HiAI 支援 TensorFlow 和 CAFFE2 模型,我們可以透過名為 cngen 的工具將 TensorFlow 或 CAFFE2 模型轉為 HiAI 引擎支援的格式。

使用者文件中提到了「線上模型」和「離線模型」這兩個術語,它們是指裝置狀態。在「離線」模型中,我們不必在裝置和雲端之間建立連接,雲端伺服器首先將模型發送到裝置,然後裝置自行執行該模型。「線上模型」表示在推理或教育訓練中,裝置和雲端之間存在連接和資料交換。

下面列出一個執行 DDK 的應用實例,可以讓讀者比較直觀地了解底層的功能。如果讀者只關注建置應用,那麼了解 SDK 就夠了。

我們從華為網站下載 DDK 的程式,可以使用 Gradle 安裝,也可以使用 Bazel 建置,編譯執行檔案的程式如下:

```
android_library(
    name = "demo_lib",
    srcs = glob([
```

```
        "java/**/*.java",
    ]),
    custom_package = "com.huawei.hiaidemo",
    manifest = "AndroidManifest.xml",
    resource_files = glob(["res/**"]),
    deps = [
        "//app/src/main/libs:android_tensorflow_inference_jar",
        "//app/src/main/libs:libhiai",
        "//app/src/main/libs:libtensorflow_inference",
        "//third_party:android_arch_core_common",
        "//third_party:android_arch_lifecycle_common",
        "//third_party:com_android_support_constraint_constraint_layout_solver",
        "//third_party:com_android_support_support_annotations",
        "@android_arch_lifecycle_runtime//aar",
        "@com_android_support_appcompat_v7//aar",
        "@com_android_support_cardview//aar",
        "@com_android_support_constraint_constraint_layout//aar",
        "@com_android_support_design//aar",
        "@com_android_support_recyclerview//aar",
        "@com_android_support_support_compat//aar",
        "@com_android_support_support_core_ui//aar",
        "@com_android_support_support_core_utils//aar",
        "@com_android_support_support_fragment//aar",
        "@com_android_support_support_v4//aar",
        "@com_android_support_support_vector_drawable//aar",
    ],
)

android_binary(
    name = "demo",
    custom_package = "com.huawei.hiaidemo",
    assets = [
        "//app/src/main/assets:InceptionV3.cambricon",
        "//app/src/main/assets:inceptionv3_cpu.pb",
```

```
        "//app/src/main/assets:labels.txt",
    ],
    assets_dir = "",
    manifest = "AndroidManifest.xml",
    manifest_values = {
        "applicationId": "com.huawei.hiaidemo",
    },
    multidex = "native",
    resource_files = glob(["res/**"]),
    deps = [
        ":demo_lib",
    ],
)
```

華為提供預先編譯的函數庫檔案作為 HiAI 的介面，下面我們重新編譯函數庫檔案：

```
package(default_visibility = [
    "//visibility:public",
])

cc_import(
  name = "libai_client.so",
  shared_library = "arm64-v8a/libai_client.so",
)
```

然後，就可以建置執行檔案了：

```
$ bazel build app/src/main:demo
```

指令執行後獲得應用文件，下面是在手機上安裝應用之後的截圖，如圖 9-10 所示。

該應用執行時期的截圖如圖 9-11 所示。

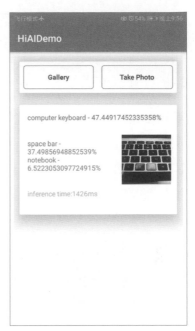

圖 9-10 華為示範應用安裝圖　　　圖 9-11 華為 HiAI 示範應用執行圖

9.2.3 簡要比較

高通 SDK 更專注於開發人員的實際的開發工作。舉例來説，參考指南文件在解釋如何有效使用其 API 方面花費了大量的文字。而華為 SDK 更專注於幫助開發人員開發不同類型的應用程式。

舉例來説，華為提供了大量的示範和 Android Studio 外掛程式，希望開發者可以用最小的資源開發出應用。高通可能需要開發者有對應的領域知識，希望開發者能夠建置應用。華為 SDK 則更多是針對普通行動開發者，比較容易上手，筆者個人希望華為的開發套件可以做得再精緻一些。

9.2.4 開放式神經網路交換格式

開放式神經網路交換格式，簡稱 ONNX，是由 Facebook 和微軟建立的社群專案。ONNX 在首頁（https://onnx.ai）上聲稱：

我們認為 AI 工具社群需要更強的互通性。許多人正在研究出色的工具，但開發人員經常被鎖定在一個架構或生態系統中。ONNX 是允許更多這些工具協作工作的第一步，允許它們共用模型。我們的目標是讓開發人員能夠為他們的專案。使用正確的工具組合。我們希望每個人都能夠儘快將人工智慧從研究變為現實，而無需工具鏈的人為摩擦。我們希望您能加入我們的使命！

除了 Facebook，微軟、亞馬遜、聯發科、AMD、華為、ARM、IBM、英特爾和高通等一些大公司也宣佈支援 ONNX。Google 和蘋果尚未就 ONNX 支援發表官方通知。

Google 和蘋果擁有自己的生態系統和人工智慧架構，因此他們可能對支援 ONNX 不有興趣。TensorFlow 社群有很多支援 ONNX 的聲音。ONNX-TensorFlow 是一個 GitHub 開放原始碼專案，它將 ONNX 支援增加到 TensorFlow 中。

現在，TensorFlow 官方還沒有正式表態對 ONNX 的支援。

10

機器學習應用架構

本章將介紹兩個機器學習的應用架構，它們都是 Google 近期推出的比較有影響力的機器學習的服務架構。

10.1 ML Kit

ML Kit 以強大且易用的方式為行動開發人員提供了 Google 的機器學習專業知識和服務。機器學習已經成為行動開發中的一種不可或缺的工具，2018 年 Google I/O 上，Google 推出了測試版 ML Kit。

ML Kit 是一款全新的 SDK，它以 Firebase 上一個功能強大但易用的軟體套件形式，將 Google 的機器學習的能力帶給廣大的行動開發者。目前，由於 Google 服務不是很容易連通，Firebase 作為 Google 服務的一部分，在中國直接使用的可能性很低。Google 也考慮到這種現狀，希望以後可以找到一個可以解決的方案。

Firebase 提供了一套行動端的服務，使用起來非常簡便。透過簡單好用的 API 提供機器學習服務，這是 Google 為降低機器學習門檻所做的努力。ML Kit 的官方網頁是 https:// developers .google.com/ml-kit。

使用機器學習對許多開發者來說可能非常困難。一般來說新的機器學習開發者需要花費大量的時間學習實現底層模型和使用架構等許多複雜操作。即使是經驗豐富的專家，修改和最佳化模型並使其在行動裝置上執行也可能是一項艱巨的工作。除機器學習的複雜性外，尋找訓練資料也可能是一個費時費力的過程。可是借助 ML Kit，讀者可以使用機器學習在 Android 和 iOS 上建置各種應用。

ML Kit 有以下三個特性：

- 對行動裝置高度最佳化。
- 依靠 Google 能力。
- 可行性和全面性。

ML Kit 可以使開發者的應用程式更具吸引力、個性化和更有用，並提供針對在行動裝置上執行而進行最佳化的解決方案。ML Kit 提供的技術是 Google 長期以來在行動裝置累積的體驗。讀者根據特定需求，可以使用開箱即用的解決方案（使用基本 API），建置在裝置上執行或在雲中執行的自訂模型。

ML Kit 透過一個簡單的介面為開發者提供裝置上的 API 和雲 API，讀者可以根據自己的需求選擇最合適的介面。裝置上的 API 可以快速處理資料，甚至可以在沒有網路連接的情況下工作；而以雲為基礎的 API 則充分利用了 Google Cloud Platform 的機器學習技術，可以提供更高的準確性。

Google 建議，如果開發者使用 Firebase，需要透過 Firebase 管理部署服務基礎架構，那麼 ML Kit 會更適合。ML Kit 提供了一組基本 API，並在 TensorFlow Lite 上層為自訂模型包裝了薄薄一層。但是，如果開發者沒有使用 Firebase 的經驗，或沒有擁有自訂模型，並希望更多地控制部署 / 服務等，Google 強烈建議開發者直接使用 TensorFlow Lite。

ML Kit 的開發是在 TensorFlow 和 TensorFlow Lite 之後，作為一個獨立的、為開發者提供機器學習解決方案的軟體服務而開發的。對於熟悉機器學習並且有機器學習部署經驗的讀者，可以直接使用協力廠商的開發軟體進行開發，對於不熟悉機器學習的讀者，或想更快地推出機器學習產品的開發者和公司，可以考慮使用類似於 ML Kit 的開發軟體。中國也有類似的為開發者提供機器學習的軟體和服務，ML Kit 的長處可以說是依靠於 Google 的服務，並且和 TensorFlow 深度整合。

如果讀者有豐富的機器學習經驗，並發現 ML Kit API 沒有涵蓋的使用案例，則可以利用 ML Kit 部署自己的 TensorFlow Lite 模型。讀者只需透過 Firebase 主控台上傳模型，ML Kit 將負責託管工作，並將它們提供給應用的使用者。這樣一來，讀者可以讓模型獨立於自己的 APK/ 軟體套件，進一步減少應用的安裝大小。

另外，由於 ML Kit 可以動態提供開發者的模型，所以我們可以更新模型，而不必重新發佈應用。

ML Kit 未來還會增加更多功能。但隨著應用的功能越來越多，它們也隨之增大，這會影響使用者從市集安裝應用的速度，並且可能讓使用者支付更多的流量費。機器學習會進一步加劇這種趨勢，因為模型的大小很容易就能達到數十百萬位元組。

因此，Google 決定開發模型壓縮功能。實際來說，Google 正在試驗一項功能，它可以讓開發者在上傳完整的 TensorFlow 模型和訓練資料後，獲得壓縮後的 TensorFlow Lite 模型。希望 Google 可以在近期推出類似的服務。

對於常見的應用使用案例，ML Kit 為開發者提供了以下範例，並快速投入產品。

- 臉部檢測。
- 文字識別。
- 條碼掃描。
- 影像標記。
- 地標識別。

我們使用 ML Kit API，只需將數據傳入 ML Kit，SDK 就會迅速獲得機器學習的結果。這裡筆者只是簡要地介紹這些實例，更詳細的技術文件可以參照 ML Kit 的官方網站。

在使用 ML Kit 之前，我們介紹一下使用的步驟。

第一步，啟動 Firbase。有關 Firebase 的實際操作內容，讀者可以參照 https://firebase.corp.google.com。如果讀者使用 Android Studio，請先安裝 Firebase 的外掛程式，使用 SDK Mananger 安裝最新的 SDK 後，在 Tool 的功能表列可以看到如圖 10-1 所示的 Firebase 外掛程式選項。

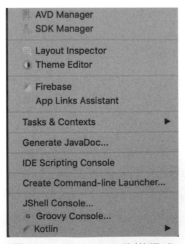

圖 10-1　Firebase 外掛程式

點擊圖 10-1 中的 Firebase 選項，在 Android Studio 裡出現 Firebase 的提示頁面，如圖 10-2 所示。

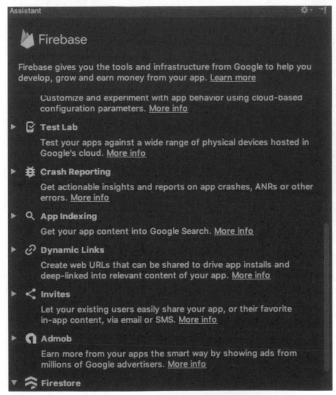

圖 10-2 Firebase 提示頁面

第二步，註冊和啟動 Firebase 專案。下面是筆者的 Firebase 首頁，如圖 10-3 所示。

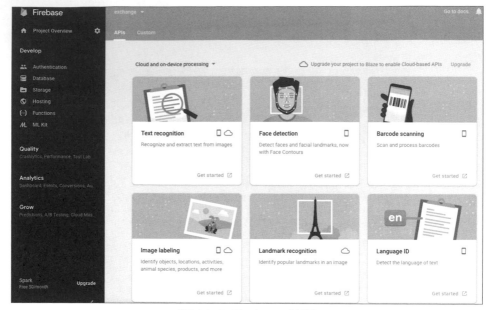

圖 10-3 Firebase 首頁

點擊圖 10-3 中右上角的 Upgrade 連結,來到啟動頁面,如圖 10-4 所示。

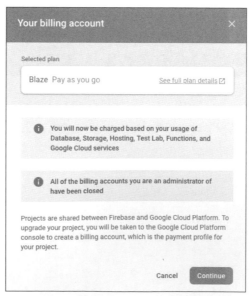

圖 10-4 Firebase 啟動頁面

這兩步完成之後，就可以開始使用和開發機器學習的應用了。

10.1.1 臉部識別（Face Detection）

使用 ML Kit 的臉部檢測 API，讀者可以檢測影像中的臉部，識別關鍵臉部特徵，並取得檢測到的臉部輪廓。透過臉部檢測，使用者可以獲得執行諸如自拍和肖像或從使用者照片產生圖示等工作所需的資訊。由於 ML Kit 可以即時執行臉部檢測，因此讀者可以在視訊聊天或回應播放機運算式的遊戲等應用程式中使用它。

臉部識別主要特性如表 10-1 所示。

表 10-1 臉部識別主要特性列表

特性	實際內容
識別並定位臉部特徵	取得檢測到的每張臉的眼睛、耳朵、臉頰、鼻子和嘴巴的座標
獲得臉部輪廓	取得檢測到的臉部及其眼睛、眉毛、嘴唇和鼻子的輪廓
識別臉部表情	確定一個人是在微笑還是閉著眼睛
追蹤視訊畫面中的面孔	取得檢測到的每個人的臉部的識別符號，此識別符號在呼叫之間保持一致，因此可以對視訊流中的特定人員執行影像處理
即時處理視訊畫面	臉部檢測在裝置上執行，並且足夠快，以滿足即時應用，例如視訊操作

在實現應用和建置程式之前，我們首先要替應用專案增加依賴。在 build. gradle 裡增加下面的依賴：

```
dependencies {
    implementation 'com.google.firebase:firebase-ml-vision:18.0.2'
    implementation 'com.google.firebase:firebase-ml-vision-face-model:17.0.2'
}
```

對於 Firebase 專案,可能要推薦使用 Gradle 作為編譯。使用 Bazel 技術上也是可能的,Google 內部也是使用 build。筆者用也用 Firebase 開發了幾個應用,但是,由於 Firebase 本身有協力廠商的依賴,build 檔案的依賴會變得很複雜,有時還會引發一些錯誤。為了快速開發,用 Gradle 是個可行的選項。

如果讀者想實現裝置端(on-device)的機器學習,需要增加下面的設定:

```
<application ...>
  <meta-data
      android:name="com.google.firebase.ml.vision.DEPENDENCIES"
android:value="ocr" />
  <!-- To use multiple models: android:value="ocr,model2,model3" -->
</application>
```

接下來講一下影像輸入標準。

要使 ML Kit 精確檢測臉部,輸入影像必須包含由足夠像素資料表示的面。一般來説要在影像中檢測的每個臉部應至少為 100×100 像素。如果要檢測臉部輪廓,ML Kit 需要更高解析度的輸入:每個臉部應至少為 200×200 像素。

如果要在即時應用程式中檢測臉部,則可能還需要考慮輸入影像的整體尺寸,以便能更快地處理較小的影像,進一步減少延遲,以較低的解析度捕捉影像並確保主體的臉部盡可能多地佔據影像。影像焦點不佳會影響精確度。如果讀者無法獲得可接受的結果,可以嘗試讓使用者重新捕捉影像。如表 10-2 所示是各種功能選項的 API 清單。

表 10-2　各種功能選項的 API 清單

API	函數	功能
FirebaseVisionFace DetectorOptions	build()	建置一個實例
FirebaseVisionFace DetectorOptions.Builder	enableTracking()	啟用臉部追蹤，在處理連續畫面時為每個臉部保持一致的 ID
FirebaseVisionFace DetectorOptions.Builder	setClassificationMode (int classificationMode)	指示是否執行其他分類器來代表「微笑」和「眨眼」等屬性
FirebaseVisionFace DetectorOptions.Builder	setContourMode (int contourMode)	設定是否檢測無輪廓或所有輪廓
FirebaseVisionFace DetectorOptions.Builder	setLandmarkMode (int landmarkMode)	設定是否檢測沒有地標或所有地標
FirebaseVisionFace DetectorOptions.Builder	setMinFaceSize (float minFaceSize)	設定所需的最小臉部大小，表示為頭部寬度與影像寬度的比例
FirebaseVisionFaceDete ctorOptions.Builder	setPerformanceMode(int performanceMode)	擴充選項，用於控制執行臉部檢測時的額外精度 / 速度權衡

此外，我們還要設定臉部識別的選項，如表 10-3 所示。

表 10-3　臉部識別的設定選項

設定項	實際內容	說明
效能模式	FAST (default)	檢測臉部時有利於速度或準確度
檢測地標模式	NO_LANDMARKS (default)	是否嘗試識別臉部「地標」：眼睛、耳朵、鼻子、臉頰、嘴巴等
檢測輪廓	NO_CONTOURS (default)	是否檢測臉部特徵的輪廓。僅針對影像中最突出的臉部檢測輪廓

設定項	實際內容	說明
臉部分類	NO_CLASSIFICATIONS (default)	是否將面孔分類為「微笑」和「睜眼」等類別
臉部最小大小	float (default: 0.1f)	要檢測的臉部的最小尺寸,相對於影像
啟動臉部追蹤	false (default)	是否為臉部分配 ID,可用於追蹤影像中的臉部

以下是設定對應的程式,

```
// 高精度地標檢測與人臉分類
FirebaseVisionFaceDetectorOptions highAccuracyOpts =
    new FirebaseVisionFaceDetectorOptions.Builder()
        .setPerformanceMode(FirebaseVisionFaceDetectorOptions.ACCURATE)
        .setLandmarkMode(FirebaseVisionFaceDetectorOptions.ALL_LANDMARKS)
        .setClassificationMode(FirebaseVisionFaceDetectorOptions.ALL_
CLASSIFICATIONS)
        .build();

// 多面即時輪廓檢測
FirebaseVisionFaceDetectorOptions realTimeOpts =
    new FirebaseVisionFaceDetectorOptions.Builder()
        .setContourMode(FirebaseVisionFaceDetectorOptions.ALL_CONTOURS)
        .build();
```

下面來看看如何實現臉部識別。要檢測影像中的臉部,請從 Bitmap、media.Image、ByteBuffer 或裝置上的檔案建立 FirebaseVisionImage 物件。然後,將 FirebaseVisionImage 物件傳遞給 FirebaseVisionFaceDetector 的 detectInImage 函數。對於臉部識別,應使用尺寸至少為 480x360 像素的影像。如果要即時識別臉部,以最低解析度捕捉畫面可以幫助減少延遲。由於相機鏡頭的轉向對識別的結果有影響,所以我們要首先獲得相機的轉向,下面是對應的程式:

```
// 取得裝置相對於其 " 本機 " 方向的目前旋轉，然後從 " 方向 " 表中尋找影像必須達到的角
度，旋轉以補償裝置的旋轉
    int deviceRotation = activity.getWindowManager().getDefaultDisplay().
getRotation();
    int rotationCompensation = ORIENTATIONS.get(deviceRotation);


// 在大多數裝置上，感測器方向為 90 度，但有些裝置是 270 度。感測器方向為 270 度，再將
影像旋轉 180 ((270+270)%360) 度
    CameraManager cameraManager = (CameraManager) context.getSystemService
(CAMERA_SERVICE);
    int sensorOrientation = cameraManager
            .getCameraCharacteristics(cameraId)
            .get(CameraCharacteristics.SENSOR_ORIENTATION);
    rotationCompensation = (rotationCompensation + sensorOrientation + 270)
% 360;
```

然後，產生一個 FirebaseVisionImage 的實例，程式如下：

```
FirebaseVisionImage image = FirebaseVisionImage.fromMediaImage
(mediaImage, rotation);
```

接著，設定中繼資料（Meta Data），告訴檢測器影像的解析度、格式和
相機的轉向，程式如下：

```
FirebaseVisionImageMetadata metadata = new FirebaseVisionImageMetadata.
Builder()
    .setWidth(480)    // 480×360 通常足以進行影像識別
    .setHeight(360)
    .setFormat(FirebaseVisionImageMetadata.IMAGE_FORMAT_NV21)
    .setRotation(rotation)
    .build();
```

另外，還要產生一個臉部檢測器的實例，程式如下：

```
FirebaseVisionFaceDetector detector = FirebaseVision.getInstance()
    .getVisionFaceDetector(options);
```

最後，呼叫檢測的功能，獲得結果 FirebaseVisionFace，程式如下：

```
Task<List<FirebaseVisionFace>> result =
    detector.detectInImage(image)
        .addOnSuccessListener(
            new OnSuccessListener<List<FirebaseVisionFace>>() {
            @Override
            public void onSuccess(List<FirebaseVisionFace>
faces) {
                    // 工作已成功地完成
                }
            })
        .addOnFailureListener(
            new OnFailureListener() {
            @Override
            public void onFailure(@NonNull Exception e) {
                    // 工作失敗，出現例外
                }
            });
```

至此，臉部識別已經完成。下面筆者透過程式來示範如何正確使用臉部
檢測結果 FirebaseVisionFace：

```
for (FirebaseVisionFace face : faces) {
    Rect bounds = face.getBoundingBox();
    float rotY = face.getHeadEulerAngleY();    // 表頭旋轉到 rotY 度
float rotZ = face.getHeadEulerAngleZ();        // 表頭側傾，旋轉角度



// 如果啟用了 Landmark 檢測（口、耳、眼、臉頰和可用鼻子）：
```

```
FirebaseVisionFaceLandmark leftEar =
        face.getLandmark(FirebaseVisionFaceLandmark.LEFT_EAR);
    if (leftEar != null) {
        FirebaseVisionPoint leftEarPos = leftEar.getPosition();
    }

    // 如果啟用輪廓檢測：
    List<FirebaseVisionPoint> leftEyeContour =
 face.getContour(FirebaseVisionFaceContour.LEFT_EYE).getPoints();
    List<FirebaseVisionPoint> upperLipBottomContour =
 face.getContour(FirebaseVisionFaceContour.UPPER_LIP_BOTTOM).getPoints();

    // 如果啟用了分類：
if (face.getSmilingProbability() != FirebaseVisionFace.UNCOMPUTED_
PROBABILITY) {
        float smileProb = face.getSmilingProbability();
    }
    if (face.getRightEyeOpenProbability() != FirebaseVisionFace.UNCOMPUTED_
PROBABILITY) {
        float rightEyeOpenProb = face.getRightEyeOpenProbability();
    }

    // 如果已啟用人臉追蹤：
    if (face.getTrackingId() != FirebaseVisionFace.INVALID_ID) {
        int id = face.getTrackingId();
    }
}
```

檢測器的主要模組 detector.detectInImage 使用了回呼函數，人臉識別是在非主執行緒上完成的。官方網站還提供了 Kotlin 和 iOS 的實例，這裡我們就不詳述了。Firebase 同時提供了 on-device 和雲端機器學習的支援，讀者可以根據需求去選擇。

10.1.2 文字識別

下面我們來實現文字識別的應用。

首先，在應用裡增加依賴：

```
dependencies {
  implementation 'com.google.firebase:firebase-ml-natural-language: 18.1.1'
  implementation 'com.google.firebase:firebase-ml-natural-language-
language-id-model:18.0.2'
}
```

然後，產生 FirebaseLanguageIdentification 的實例，並呼叫函數去識別文字，實現程式如下：

```
FirebaseLanguageIdentification languageIdentifier =
    FirebaseNaturalLanguage.getInstance().getLanguageIdentification();
languageIdentifier.identifyLanguage(text)
  .addOnSuccessListener(
    new OnSuccessListener<String>() {
      @Override
      public void onSuccess(@Nullable String languageCode) {
        if (languageCode != "und") {
          Log.i(TAG, "Language: " + languageCode);
        } else {
          Log.i(TAG, "Can't identify language.");
        }
      }
    })
  .addOnFailureListener(
    new OnFailureListener() {
      @Override
      public void onFailure(@NonNull Exception e) {
```

```
        // 無法載入模型或其他內部錯誤
    }
});
```

10.1.3 條碼識別

下面，我們來實現一個條碼識別的應用。同樣，要先增加依賴，再產生
實例，接著呼叫功能去實現條碼，程式如下：

```
FirebaseVisionBarcodeDetectorOptions options =
    new FirebaseVisionBarcodeDetectorOptions.Builder()
    .setBarcodeFormats(
            FirebaseVisionBarcode.FORMAT_QR_CODE,
            FirebaseVisionBarcode.FORMAT_AZTEC)
    .build();
```

上面我們介紹了一些基本應用的實現，這些已經可以覆蓋大部分人工智
慧的應用。此外，我們能感受到應用所需的程式量非常少，更重要的
是，它可以幫助我們省去複雜的機器學習操作和管理的步驟，而專注於
商業邏輯的開發。應用還可以在裝置端學習和雲端學習之間進行轉換，
大幅加強了應用的適用場景。

在實際應用中，ML Kit 獲得了不錯的反響。行動端的推測是以
TensorFlow Lite 為基礎的，所以計算的效果和效率還是不錯的。各個應
用的模型 Google 內部也在使用，應該能提供比較好的使用者體驗。

今後，ML Kit 還會提供更多的模型。Google 團隊也在加強 TensorFlow
Lite 的能力，以及提供更好的行動端體驗，希望 ML Kit 會有一個更廣闊
的應用前景。

10.2 聯合學習（Federated Learning）

Google2017 年發佈了聯合學習（Federated Learning）的論文，網址是 https://arxiv.org/ abs/1602.05629，讀者有興趣可以找原文閱讀。同時，Google 也發表了一篇部落格，網址是 https://ai.googleblog.com/2017/04/federated-learning-collaborative.html，算是給自己做個宣傳，同時也解釋一下聯合學習。

在論文發表之前，Google 已經開始了這方面的研究和產品開發，這個技術被應用到 Google 鍵盤（Gboard）上。這篇論文既是理論上的發表，也是工程實作的歸納。聯合學習的基本概念我們可以透過圖 10-5 來說明。

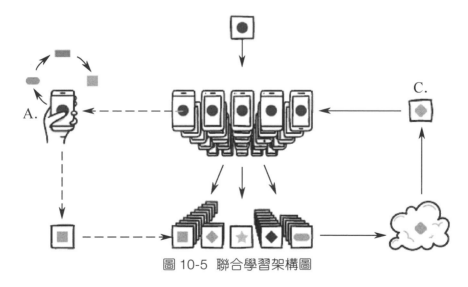

圖 10-5 聯合學習架構圖

聯合學習是利用數量龐大的行動裝置進行機器學習的一種方法。下面介紹一下聯合學習是如何解決資料的隱私問題的。

資料的隱私是一個很重要的問題。從行動端取得使用者資料，似乎是最簡單最直接的方法。可是在美國，例如在 Google 內部，資料保密和隱私是非常敏感的問題，近幾年來，資料隱私變得越來重要。沒有使用者的同意，任何產品和服務是不能獲得使用者資料的。

中國的企業很重視使用者的隱私。聯合學習是用數學的方法解決了這個問題。實際地講，就是行動端不會向後台或協力廠商傳送使用者資料，使用者資料會永遠留在裝置裡，機器學習的過程完全是以行動裝置為基礎的（on-device）。

機器學習會把訓練後的模型傳到後端，後端把行動端傳送來的模類型資料進行整理和再處理，包含再訓練和評價，進而產生一個最佳化的模型，這個模型會被再次發送到行動端。為了實現這個目的，還有幾個問題需要解決。行動端的每個裝置都是相對獨立的，這和以雲端為基礎的分散式訓練有很大的不同。這表示，行動端的機器學習訓練是不可預期的，例如訓練發生的時間不確定，雖然我們可以透過程式設定訓練的頻率和開始時間，但是使用者隨時可以關機或改變裝置狀態而引起一些訓練時間的改變。

機器學習的訓練在行動裝置上執行是非常具有挑戰性的，有時訓練的結果和過程都存在很大的不確定性。聯合學習需要裝置端把訓練後的結果傳回後端，但是，由於通訊和連接的問題，很難保障裝置按照預定的時間表把結果傳回來。所以，後端要對接收到的模型參數進行處理。

行動端把訓練後的模型參數傳到後台後，即使模型的參數洩露，透過對這些參數的分析，我們還是能獲得模型的大概資訊，透過反向過程，甚至可以推出使用者資料。聯合學習使用了安全聚合協定（Secure Aggregation Protocol ）來處理這個問題。

聯合學習的原理是由一系列論文組成的，這些論文有關行動端機器學習的各方面。雖然有了這些理論，實作起來卻一點也不輕鬆。主要原因是，在行動端上進行機器學習和在資料中心或以桌上型電腦為主進行機器學習有很大的不同，可能要分成幾個小的團隊來集體開發這套系統，其中包含一個負責行動開發的團隊，一個負責後台的團隊。整個系統還要包含監測，以及對這個系統的管理等，工程量還是很大的。

聯合學習的流程如圖 10-6 所示。

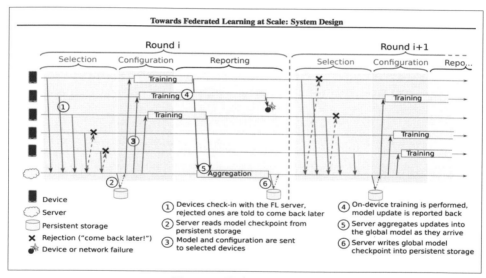

圖 10-6　聯合學習的流程

聯合學習的基本演算法如圖 10-7 所示。

Algorithm 1 `FederatedAveraging` targeting updates from K clients per round.

Server executes:
  initialize w_0
  for each round $t = 1, 2, \ldots$ **do**
    Select $1.3K$ eligible clients to compute updates
    Wait for updates from K clients (indexed $1, \ldots, K$)
    $(\Delta^k, n^k) = \text{ClientUpdate}(w)$ from client $k \in [K]$.
    $\bar{w}_t = \sum_k \Delta^k$ *// Sum of weighted updates*
    $\bar{n}_t = \sum_k n^k$ *// Sum of weights*
    $\Delta_t = \Delta_t^k / \bar{n}_t$ *// Average update*
    $w_{t+1} \leftarrow w_t + \Delta_t$

ClientUpdate(w)**:**
  $\mathcal{B} \leftarrow$ (local data divided into minibatches)
  $n \leftarrow |\mathcal{B}|$ *// Update weight*
  $w_{\text{init}} \leftarrow w$
  for batch $b \in \mathcal{B}$ **do**
    $w \leftarrow w - \eta \nabla \ell(w; b)$
  $\Delta \leftarrow n \cdot (w - w_{\text{init}})$ *// Weighted update*
  // Note Δ is more amenable to compression than w
  return (Δ, n) to server

圖 10-7 聯合學習演算法

以行動裝置為基礎的
機器學習的未來

本章我們將了解機器學習最新的一些進展和趨勢。

首先來看一下 TensorFlow 的最新動態。

11.1 TensorFlow 2.0 和路線圖

TensorFlow 本身是一個快速發展的開放原始碼社群支援的專案。接下來簡單介紹一下 TensorFlow 2.0 和未來的發展趨勢。

TensorFlow 2.0 將是一個重要的里程碑，這個版本的重點是便利性。以前 TensorFlow 可能只有有經驗的開發者才能使用，TensorFlow 2.0 希望能透過降低使用門檻，讓更多的開發者使用 TensorFlow，未來能看到更多的人工智慧使用場景。

TensorFlow 2.0 的主要特點如下：

- 使用 Keras 快速開發模型。

- 立即執行（Eager Execution）是 2.0 版本的核心功能。這使 TensorFlow 更易於學習和應用。

- 使用跨平台的、更可靠的產品模型發佈。

- 透過交換格式的標準化和 API 的一致性，支援更多平台和語言，並改善元件間的相容性。

- 為研究人員提供更有效的實驗平台。

- 具有更簡單的 API，刪除過時和重複的 API。

在過去幾年中，TensorFlow 增加了許多元件。透過 TensorFlow 2.0，這些元件將被包裝成一個綜合平台，支援從訓練到部署的完整的機器學習工作流程。讓我們看一下 TensorFlow 2.0 的新架構，在此使用簡化的概念圖，如圖 11-1 所示。

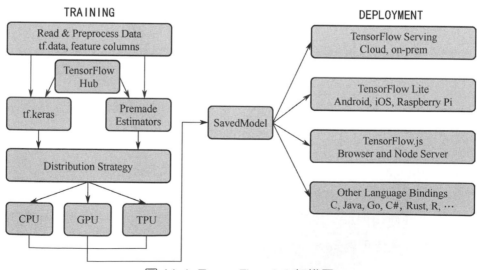

圖 11-1 TensorFlow 2.0 架構圖

11.1.1 更簡單的開發模型

在 TensorFlow 2.0 中，Keras 會成為一個使用者人性化的機器學習進階 API 的標準，並將成為用於建置和訓練模型最重要的進階核心 API。 Keras API 讓開發者可以輕鬆開始使用 TensorFlow。重要的是，Keras 提供了一些模型建立 API，讓開發者可以為開發專案選擇合適的抽象等級。TensorFlow 2.0 的功能還包含 Eager Execution、快速反覆運算和直接偵錯，以及用於建置可擴充輸入管線的 tf.data。

11.1.2 更可靠的跨平台的模型發佈

TensorFlow 2.0 將相當大加強跨平台的相容性，使模型從開發到產品發佈更加流暢。2.0 版本的跨平台相關的要點如下：

- 使用 TensorFlow Serving。現在支援 HTTP/REST 和 gRPC。
- 使用 TenforFlow Lite。對於行動裝置、IoT 和邊緣裝置，使用新的 TensorFlow Lite。
- 使用 TensorFlow js。網頁端使用 TensorFlow js。

11.1.3 TensorFlow Lite

這裡我們重點看一下 TensorFlow Lite。

TensorFlow Lite 是針對行動裝置、嵌入裝置的預設架構。我們已經看到 Google 內部團體也在大量採用 TensorFlow Lite。同時，很多開發者也在開發以 TensorFlow Lite 為基礎的產品，而且市場上也出現了很多成功的產品。TensorFlow Lite 和 Edge TPU 對軟體和硬體搭檔會有很好的應用前景，即使作為獨立的套件，它們也會各自獨當一面。

但是，客觀地看，TensorFlow Lite 距離 TensorFlow 的成熟度還有一段距離，而且也面臨很大的挑戰。

今後 TensorFlow Lite 將要注重發展的方向如下：

- 增加 TensorFlow Lite 中支援的運算元的覆蓋範圍。
- 更容易轉換訓練過的 TensorFlow 圖，以便在 TensorFlow Lite 上使用。
- 加強用於移動模型最佳化的工具。
- 擴充對 Edge TPU、TPU AIY 開發板的支援。
- 更好的文件和教學。

我們期待更好、更好用和更高效的機器學習架構。

11.1.4 TensorFlow 1.0 和 TensorFlow 2.0 的不同

除了以上 TensorFlow 2.0 的新特性，2.0 版本還清除了一些不合適和過時的 API，並使 API 更加統一。其中一個大的變化是，tf.contrib 將被移除！

Contrib 在 1.0 版本是一個實驗產品的集合，由於 TensorFlow 快速普及和被大量採用，Contrib 變得越來越大。從 2.0 版本開始，只有核心功能會加到主幹，其他特性和分支會獨立出來，盡可能將 tf.contrib 中的大型專案移到單獨的程式庫中。TensorFlowLite 已經在 2018 年底從 Contrib 裡移了出來。

11.2 人工智慧的發展方向

本書的最後，筆者想討論一下人工智慧的未來發展方向。既然
TensorFlow 是由 Google 發起的，筆者就整理一下 Google 對人工智慧未
來發展的一些公開的文章和論點，也會有一些個人看法。

最近幾年，隨著機器學習和深度學習的發展，人工智慧已經深入我們生
活的各方面，有些已經成為生活中不可缺少的一部分。人工智慧以後的
發展對我們的生活將有更深遠的影響，接下來從三個方面進行探討：

- 加強人工智慧的可解釋性。
- 貢獻社會。
- 改善生活。

11.2.1 加強人工智慧的可解釋性

如何能夠讓人工智慧更加透明，讓我們更進一步地了解機器學習的結
果，是以後人工智慧發展的重要方向。就像今天我們仍然無法完全了解
人腦的活動和人類的思考方式，機器學習的基本原理也是在模仿人類的
思考方式。

許多深度學習演算法自誕生起就一直被人們視作「神秘黑匣」，這是因為
就連它們的發明者也難以準確表達輸入和輸出之間究竟發生了什麼。如
果我們繼續把人工智慧當作「神秘黑匣」，那麼我們就不能期望獲得使用
者和社會的信任，因為信任來自了解。

對於傳統軟體，我們可以透過逐行檢查原始程式碼來揭示其中的邏輯，
但神經網路是一個透過數千個乃至數百萬個訓練範例而形成的密集連接

網路，所以很難用已有的方式去了解裡面的邏輯關係。這也造成了人們對機器學習的不信任和更多的神秘感。

現在的機器學習，快速地還是非常依賴資料的。透過大量資料的訓練讓機器去了解事物的本質，這是一種有效的方式，也是現在人工智慧的主流方式，這和人類的認知方式也很相像。但是，當我們反問自己，我們是否真正了解了這個過程的實際意義？我們真正了解了結果的含義嗎？

可能我們並不能完美地回答自己。

科學的發現和工程的進展是一個螺旋上升的過程，被證明是有效的事物，會更加促進我們去了解這個事物的本質。如果我們對人類學習有了更加深入的了解，也許會使機器學習上一個新的台階。

隨著最佳做法的建立，有效而成熟的工具越來越多，再加上大家都在努力從開發週期開始就盡可能取得可解釋的結果，這方面的研究正在不斷取得進展。

業界已經在實作中思考可解釋性的問題。舉例來說，在影像分類領域，Google 最近的研究示範了一種「人性化表示」的概念，例如影像分類器能夠根據對人類使用者最有意義的特點來清楚表達其推理過程。一個相關的實例是，影像分類器可能將影像歸類為「斑馬」，部分原因在於影像中的「條紋」特徵較明顯，而「小數點花紋」特徵相對不夠明顯。

實際上，研究人員正在試驗將這種方法應用於糖尿病視網膜病變的診斷，它可以使輸出結果更加透明。當專家不同意模型的推斷時，甚至允許對模型進行調整。

11.2.2 貢獻社會

如何能夠確定機器學習的模型對社會中的每一個個體都有貢獻，也是以後研究的方向。機器學習越來越成為社會的基礎服務和技術架構，我們每天都能接觸到機器學習的一些應用。但當我們享受到人工智慧帶來的服務的時候，我們也要保障身處社會中的每一個人都能受惠於新技術。

當正常人能夠使用人工智慧服務的時候，我們也要保障盲人和聾啞人等也能夠同樣使用新技術，分享新技術帶來的福利。在 Google，每個新的產品發佈都有一個嚴格和標準的審查流程，審查中的一項就是這個產品是否可以被有障礙的人群使用，對有障礙人群會不會產生不好的效果。

我們設計的產品，都有自己的特定人群和場景，這無可厚非，但是如果我們能多考慮一些特殊的場景，為兒童、有障礙人士、女士設計出更多符合他們生活習慣的產品，也是技術對這個社會做出的貢獻。

另一個角度是，即使我們有意識地為全體人群和使用者考慮設計產品，但是由於設計的失誤和欠缺，也會導致最後的產品產生偏差。舉例來說，我們在收集資料的時候，有沒有考慮資料的全面性。如果我們從和我們有共同生活樣式的人群裡收集資料並設計模型，這個模型很難適用於全體人群。這是非常明顯的實例，似乎沒人會犯這個錯誤。但是，如果我們重新設計模型，從有限的資料開始訓練模型的時候，一定要考慮模型的偏差，避免相似的錯誤。

此外，由於我們自己存在偏見，因此即使是如實收集資料，也可能把這種偏見表現出來。舉例來說，大量的歷史文字經常用於訓練有關自然語言處理或翻譯的機器學習模型，如不修正，可能會使某種有害的成見持續下去。

有研究嘗試用清晰度量化的方式來反映這一現象。實際研究證明，統計語言模型能夠非常輕鬆地「學習」關於性別的過時假設。例如「醫生」是「男性」，「護士」是「女性」。與此相似的偏見問題在人群族裔方面也有表現。業界正在多個領域解決這些問題，其中以感知領域最為重要。為了促進人們更廣泛地了解公平對於機器學習等技術的必要性，業界在教育界投入了大量資源，希望在教育速成課程中推廣和普及對公平性的了解。

當然，業界一直致力於為開發者提供值得信賴的工具，在去除偏見方面也是一樣。首先從文件開始，例如增強機器學習指南文件，Google 將此種指南整合在 AutoML 中，並擴充到類似 TensorFlow Model Analysis（TFMA）和 What-If 等的工具裡。該指南為開發者提供所需的分析資料，使開發者確信他們的模型會公平對待所有使用者。TFMA 可以輕鬆地將模型在使用者群眾的不同環境、特徵和子集下的效能表現視覺化，而 What-If 支援開發者執行反設事實，闡明關鍵特徵（例如指定使用者的人口屬性）逆轉時可能會發生的情況。這兩個工具都可以提供沉浸式的互動方法，用於詳細探索機器學習行為，幫助我們識別公平性和代表性方面的失誤。

對於目前採取的這些做法，筆者相信這些知識和正在開發的工具具有深遠的意義，必將促進人工智慧技術的公平性。但沒有一家公司能夠獨自解決如此複雜的問題。這場對抗偏見的戰鬥將是一次全社會的集體行動，由許多利益相關的人和群眾投入，共同推動。

11.2.3 改善生活

我們也希望，科學家和工程師能夠負責任地利用人工智慧和自動化技術，來確保我們能更進一步地改善生活，並為未來做好準備。自從人工智慧被廣泛應用之後，我們常常聽到的是以後哪些人群、哪些企業會被人工智慧所代替，這並不是危言聳聽，而是正在發生的事實。

不過，筆者認為，人工智慧的未來並非一場零和遊戲，或是對人類的毀滅。最近的一份報告也顯示，很多企業的高管認為，透過人工智慧和人類智慧相結合，人工智慧將助推人類和機器協作工作，發揮更強大的作用。

另外，我們要知道，工作很少會是單一、簡單的。大多數工作都是由無數不同工作組成的，從高度創新到重複性工作，每一項工作都會在特定程度上受到自動化技術的影響。我們希望這種改變能推動社會向好的方向發展，而非向毀滅的方向發展。

人類的每次技術革命都會產生陣痛，在這次的人工智慧變革中，我們也希望技術能減少陣痛對社會的衝擊。舉例來說，透過人工智慧技術，對兒童進行更高效的教育，對未來從事可被取代工種的人群提供個性化的教育訓練等。

然而，某些類別的工作面對的變化要比其他工作更加劇烈，並且要做出更多努力才能緩和這種轉變。但這並不是人類第一次面臨這樣的挑戰和困難。人類歷史上第一次和第二次工業革命給社會帶來的衝擊都非常極大。

從資訊革命以後，人類的技術進步和革新的步伐並沒有停止，而是越來越快。有些企業消失了，又有些新的企業誕生了。我們希望透過技術改善生活，同時，正視技術淘汰，正確面對越來越快的改變。在這種改變中，技術應該是推動社會、改善生活的正面角色，而非人類的敵人。